MANUAL FOR OPERATIONAL AMPLIFIER USERS

MANUAL FOR OPERATIONAL AMPLIFIER USERS

John D. Lenk

Consulting Technical Writer

RESTON PUBLISHING COMPANY, INC.

A Prentice-Hall Company

Reston Virginia

Library of Congress Cataloging in Publication Data

Lenk, John D
    Manual for operational amplifier users.

    1. Operational amplifiers — Handbooks, manuals, etc.
I. Title
TK7871.58.06L46   621.3815'35   75–33964
ISBN 0–87909–477–X

© 1976 by
Reston Publishing Company, Inc.
A Prentice-Hall Company
Reston, Virginia 22090

10   9   8   7   6   5   4   3   2   1

Printed in the United States of America.

This book is dedicated to my wife **Irene** for all of her patience.

CONTENTS

PREFACE

As the title implies, this manual is written for *users* of operational amplifiers (op-amps), rather than for designers of op-amps. The book, therefore, is written on the basis of using existing commercial op-amps to solve design and application problems. Typical users include design specialists who want to include op-amps or op-amp circuit applications in their systems, or technicians who must service equipment containing op-amps. Other groups that can make good use of this approach to op-amps are the experimenters and hobbyists.

Most commercial op-amps available today are in integrated circuit (IC) form. Thus, this manual assumes that the reader is already familiar with IC basics (packaging, temperature considerations, internal construction, power supplies, etc.). Such information is discussed fully in the author's *Manual For Integrated Circuit Users* (Reston Publishing Company, Reston, Virginia, 1973) and *Manual For MOS Users* (Reston, 1975). However, no direct reference is required to either book from this manual. Also, the application information contained in this manual can be used with any type of op-amp, IC or discrete component. Much of the material can even be used with vacuum-tube op-amps.

The approach found in this manual serves a threefold purpose: (1) to acquaint the readers with op-amps, in general, so that the users can select commercial units to meet their particular circuit requirements, (2) to provide users with basic op-amp characteristics so that they may specify a logical set of design requirements where op-amps are being manufactured to order, and (3) to show readers the many other uses for existing op-amps, not always found on the manufacturer's datasheet.

The manual assumes that the reader is already familiar with basic electronics, including solid state, but has little or no knowledge of op-amps. For this reason, Chapter 1 provides an introduction to op-amps. Such topics as typical op-amp configurations, circuit, datasheet characteristics and phase compensation schemes are covered.

With basics out of the way, Chapter 2 discusses linear applications for op-amps. These linear applications include those cases in which inputs and output are essentially sinewaves, even though modified by the op-amp or the associated circuit.

Chapter 3 is devoted entirely to the Operational Transconductance Amplifier (OTA). Various commercial versions of the OTA are used as a substitute for the conventional op-amp, and in applications in which the OTA is superior to an op-amp.

Chapter 4 discusses nonlinear applications for op-amps. These non-linear applications include those cases in which inputs and outputs are essentially non-sinewave, or in which the output is drastically modified by the op-amp. Other applications found in Chapter 4 also include those circuits in which the op-amp is used to generate nonlinear signals (such as square waves, pulses, integrated waves, etc.).

Chapter 5 provides detailed test data for op-amps. Often, experimenters must work with op-amps on which complete data are not available. Under these circumstances, it is necessary to test the op-amp under simulated operating conditions. Likewise, it is often necessary to test commercial op-amps against datasheet specifications. Typical test information is given for all commonly used op-amp datasheet specifications.

With any op-amp, it is possible to apply certain approximations or guidelines for the selection of external component values. These guide-lines can then be stated in basic equations, requiring only simple arithme-tic for the solutions.

This manual starts with guidelines for the selection of external com-ponents on a trial-value basis, assuming a specified goal and a given set of conditions. The manual concentrates on simple, practical approaches to op-amp use. Op-amp circuit analysis and theory is included only where necessary to make intelligent use of the op-amp.

The values of external components used with op-amps depend upon op-amp characteristics, available power sources, the desired performance (voltage amplifications, stability, etc.) and existing circuit conditions (input/output impedances, input signal amplitude, etc.). The op-amp characteristics are to be found in the manufacturer's datasheets (or by actual test).

With op-amp characteristics established, the overall system character-istics can then be determined, based on a reasonable expectation of the op-amp capabilities. Often the final circuit is a result of many tradeoffs between desired performance and available characteristics. This manual discusses the problem of tradeoffs from a simplified, practical approach.

Since the manual does not require advanced math or theoretical study, it is ideal for the experimenter. On the other hand, the manual is suited to schools where the basic teaching approach is circuit analysis, and a great desire exists for practical design.

The author has received much help from many organizations and individuals prominent in the field of op-amp technology. He wishes to thank them all, particularly Motorola Semiconductor Products, Inc., and the Solid State Division of Radio Corporation of America (RCA). The author also wishes to express his appreciation to Mr. Joseph A. Labok of Los Angeles Valley College for his help and encouragement.

John D. Lenk

1. INTRODUCTION TO
OPERATIONAL AMPLIFIERS

An operational amplifier (or op-amp) is basically a very high-gain, direct-coupled amplifier that uses *feedback* for control of response characteristics. The designation op-amp was originally used for a series of high-performance direct-coupled amplifiers that formed a basic part of analog computers. These amplifiers were used to perform mathematical operations applicable to analog computation (summation, scaling, subtraction, integration, etc.).

Today, the availability of inexpensive op-amps (particularly in integrated circuit form) has made the packaged op-amp useful as a replacement for any low-frequency amplifier. For example, the same op-amp used for mathematical operations may be adapted to provide either the broad, flat frequency-gain response required of video amplifiers or the peaked responses required of various types of shaping amplifiers.

The configuration most commonly used for op-amps is a cascade of two *differential amplifier circuits* together with an appropriate output stage. The cascaded differential amplifier stages not only fulfill the op-amp requirement for a high-gain, direct-coupled amplifier circuit, but also provide significant advantages with respect to application.

From an applications standpoint, an op-amp that has a differential input is much more versatile than a single-input type. This increased versatility results from greater flexibility in selection of the feedback configuration. With a single-input op-amp, only an *inverting feedback* configuration can be employed. When differential inputs are used, the feedback configuration may be either an inverting type or a *non-inverting type,* which depends on the *common-mode rejection* for the negative feedback. The type of feedback affects the characteristics of an op-amp, and the two types tend to complement each other. Because the char-

1

acteristics of each type are required equally often, the differential-input op-amp is twice as versatile as the single input type.

The capabilities and limitations of op-amps are firmly defined by a few simple equations and rules, which are based on a certain set of criteria that an op-amp must meet. Effective use of these simple relationships, however, requires knowledge of the conditions under which each is applicable so that errors that may result from various approximations are held to a minimum.

In this chapter we shall discuss the basic op-amp. (The many applications for op-amps are discussed in Chapters 2 through 4.) Here, we shall concentrate on typical op-amp circuits, how to interpret op-amp datasheets, and design considerations for frequency response and gain. Frequency instabilities in the op-amp and the methods used to prevent them are also discussed.

Most of the basic design information for a particular op-amp can be obtained from the datasheet or other catalog information. Likewise, a typical datasheet may describe a few specific applications for the op-amp. However, op-amp datasheets (particularly IC op-amp datasheets) generally have two weak points. First, they do not show how the listed parameters relate to design problems. Second, they do not describe the great variety of applications for which a basic op-amp can be used.

In any event, it is always necessary to interpret op-amp data, both from published material and actual test. Each manufacturer has its own system of datasheets and related technical information. It is impractical to discuss all data systems here. Instead, we shall discuss the typical information found on op-amp datasheets and see how this information affects design and use.

1-1. OPERATIONAL AMPLIFIER CIRCUITS

Operational amplifiers generally use several differential stages in cascade to provide common mode rejection and high gain. Differential amplifiers require both positive and negative power supplies. Since a differential amplifier has two inputs, it provides phase inversion for degenerative feedback, and can be connected to provide either in-phase or out-of-phase amplification.

A conventional op-amp requires that the output be fed back to the input through a resistance or impedance. The output is fed back to the negative or inverting input so as to produce degenerative feedback (to provide the desired gain and frequency response). As in any amplifier, the signal shifts in phase as it passes from input to output. This phase shift is dependent upon frequency. When the phase shift approaches

180° it adds to (or cancels out) the 180° feedback phase shift. Thus, the feedback is in phase with the input (or nearly so) and will cause the amplifier to oscillate. This condition of phase shift with increased frequency limits the bandwidth of an op-amp. The condition can be compensated by the addition of a phase shift network (usually an RC circuit, but sometimes a single capacitor).

Phase shift problems and frequency considerations are discussed in Sec. 1-2. Before going into these subjects it is necessary to understand the operation of basic op-amp circuits. We shall start with a review of the basic differential amplifier, and define the characteristics associated with that circuit.

1-1.1 Differential amplifier circuits

The differential amplifier is similar to an emitter-coupled amplifier, except that the two output signals are the result of a *signal difference* between the two inputs. In a theoretical differential amplifier, no output is produced when the signals at the input are identical. That is, an output is produced *only when there is a difference* in signals at the input. Signals common to both inputs are known as *common-mode signals*. The ability of a differential amplifier to prevent conversion of a common-mode signal into a difference signal (which produces an output) is expressed by the *common-mode rejection ratio (CMR or CMRR)*.

One of the main reasons for use of a differential amplifier as the input stage of an op-amp is that the op-amp may be operated in the presence of radiated signals (power line radiation, stray signals from generators, etc.). Leads connected to the input terminals will pick up these radiated signals, even when the leads are shielded. If a single-ended input is used, the undesired signals will be picked up and amplified along with the desired signal input. If the op-amp has a differential input, both leads will pick up the same radiated signals at the same time. Since there is no difference between the radiated signals at the two inputs, there is no amplification of the undesired inputs.

Figure 1-1 is the schematic of a basic differential amplifier. The circuit responds differently to common-mode signals than it does to a single-ended signal.

A common-mode signal (like power-line pickup) drives both bases in phase with equal-amplitude voltages, and the circuit behaves as though the transistors are in parallel to cancel the output. In effect, one transistor cancels the other.

Normal signals are applied to either of the bases (Q_1 or Q_2). The *inverting input* is applied to the base of transistor Q_2, and the *non-inverting*

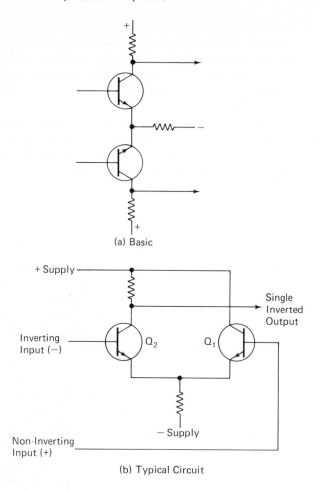

(a) Basic

(b) Typical Circuit

Fig. 1-1. Basic differential amplifier circuits.

input is applied to the base of Q_1. With a signal applied only to the inverting input, and the non-inverting input grounded, the output is an amplified and inverted version of the input. For example, if the input is a positive pulse, the output is a negative pulse. If the non-inverting input is used with the inverting input grounded, the output is an amplified version of the input (without inversion).

The emitter resistor introduces emitter feedback to both transistors simultaneously. This reduces the common-mode signal gain without reducing the differential signal gain in the same proportion.

Figure 1-2 is the schematic of a more practical differential amplifier. This circuit is typical of those found at the input of an op-amp. The circuit is basically a single-stage differential amplifier (Q_2 and Q_4) with input emitter followers (Q_1 and Q_5) and *constant-current source* Q_3 in the emitter-coupled network. Note that the single emitter resistor of the circuit in Fig. 1-1 is replaced by Q_3 and its associated circuits in the circuit of Fig. 1-2.

The use of a transistor such as Q_3 is typical for most differential amplifiers found in op-amps. The circuit of transistor Q_3 is known as a *temperature-compensated constant-current source.* All current for the differential amplifier is fed through Q_3 (an NPN) connected between the emitters of the differential amplifier and V_{EE} (the negative power supply). If there is an increase in current, a larger voltage is developed across the current-source Q_3 emitter resistor. This larger voltage acts to reverse bias the base-emitter junction, thus reducing current through Q_3. Since all current for the differential amplifier is passed through Q_3, current to the amplifier is also reduced. If there is a decrease in current, the opposite occurs, and the amplifier current increases. Thus, the differential amplifier is maintained at a constant current level.

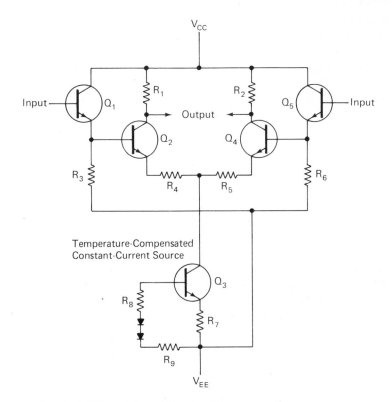

Fig. 1-2. Practical differential amplifier circuit.

Transistor Q_3 is also *temperature compensated* by the diodes connected in the base-emitter bias network. These diodes have the same (approximate) temperature characteristics as the base-emitter junction, and offset any change in Q_3 base-emitter current flow that results from temperature change.

1-1.2 Common-mode definitions

The terms *common mode* and *common-mode rejection* are used frequently in op-amp applications. All manufacturers do not agree on the exact definition of common-mode rejection.

One manufacturer defines common-mode rejection (*CMR*, or sometimes listed as CM_{rej}), or the common-mode rejection ratio (*CMRR*), as the ratio of differential gain (usually large) to common-mode gain (usually a fraction). That is, the amplifier may have a large gain of differential signals (different signals at each input terminal, or one input terminal grounded and the opposite input terminal with a signal), but little gain (or possibly a loss) of common-mode signals (same signal at both terminals).

Another manufacturer defines *CMR* as the relationship of *change* in output voltage to the *change* in the input common-mode voltage producing it, divided by the open-loop gain (amplifier gain without feedback).

For example, using the latter definition, assume that the common-mode input (applied to both terminals simultaneously) is 1V, the resultant measured output is 1 mV, and the open-loop gain is 100. The *CMR* is then:

$$\frac{(\text{output/input})}{\text{open-loop gain}} = CMR$$

$$\frac{(0.001/1)}{100} = 100{,}000 = 100 \text{ dB}$$

Another method used to calculate *CMR* is to divide the output signal by the open-loop gain to find an *equivalent differential input signal*. Then the common-mode input signal is divided by this equivalent differential input signal. Using the same figures as in the previous *CMR* calculation:

$$\frac{\text{output signal}}{\text{open-loop gain}} = \text{equivalent differential input signal}$$

$$\frac{0.001\text{V}}{100} = 0.00001$$

$$\frac{1V}{0.00001} = 100,000 = 100 \text{ dB}$$

No matter what basis is used for calculation, *CMR* is an indication of the *degree of circuit balance* of the differential stages, since common-mode input signals should be amplified identically in both halves of the circuit. A large output for a given common-mode input is an indication of large unbalance or poor *CMR*. If there is an unbalance, a common-mode signal becomes a differential signal after it passes the first stage.

As with amplifier gain, *CMR* usually decreases as frequency increases. However, as a rule of thumb, the *CMR* of any differential amplifier should be at least 20 dB *greater* than the open-loop gain at any frequency (within the limits of the op-amp).

1-1.3 Floating inputs and ground currents

Since a differential amplifier is sensitive only to the difference between two input signals (in theory), the signal source need not be grounded and can be *floating*. Thus, op-amps with differential inputs are often used in applications where the signal source is from a bridge (such as a bridge-type transducer) and the signal source is grounded.

A floating-input circuit can create problems. When the input is floating, cable shielding between the op-amp and signal source may be connected to chassis ground rather than to signal ground. However, both ac and dc voltages can exist between two widely separated earth grounds, causing current to flow. (Such currents are known as *ground currents,* and the circuits producing the current flow are known as *ground loops.*) The condition is shown in Fig. 1-3 where a bridge-type transducer is used with a differential op-amp.

Note that the signal source is connected to the transducer earth ground (local ground or physical ground, as it may sometimes be called). This ground point is connected to the op-amp ground through the cable shielding. The op-amp ground is connected to one of the differential inputs through the internal capacitance (represented as C_{INT}) of the op-amp, even though there may be no dc connection between ground and the input terminal of the floating-input op-amp.

The same differential input terminal is connected to the signal source through the signal leads and the transducer elements (bridge resistors in this case). Thus, the ac ground currents are mixed with the signal currents. This can result in an unbalance of the differential amplifier. Also, radiated signals picked up by the shield appear as undesired dif-

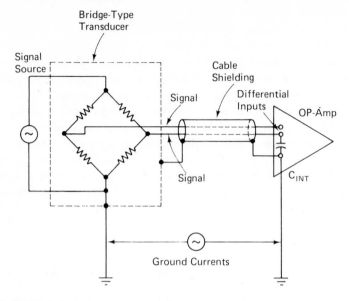

Fig. 1-3. Bridge-type transducer used with differential amplifier.

ferential signals, rather than common-mode signals, and produce an undesired output.

One method used to minimize this condition is shown in Fig. 1-4. Here,

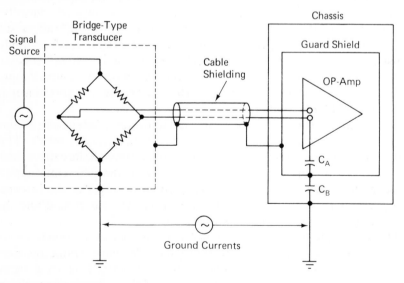

Fig. 1-4. Floated-input op-amp with guard-shield to reduce capacitance between signal leads and ground.

a guard shield is placed around the input circuits of the differential amplifier. This not only shields the differential amplifier from radiated signals but also provides an electrostatic shield to break the internal capacitance C_{INT} into two series capacitances C_A and C_B. A much higher impedance is then presented to the flow of ac ground signals. This type of op-amp is termed a *floated-input* and *guarded* op-amp.

1-1.4 Balanced differential IC op-amp

The most common type of IC op-amp uses a *balanced differential amplifier* circuit. Figure 1-5 shows a typical circuit. The complete circuit shown is contained in a ceramic flat pack (less than one-half inch square), which has 14 terminal leads. Note that no internal capacitors are used. Only resistors and transistors are used.

The basic purpose of such a circuit is to produce an output signal that is linearly proportional to the difference between two signals applied to the input. The circuit shown will provide an overall loop gain of approximately 2500.

The following is a brief description of the circuit shown in Fig. 1-5. As illustrated, the op-amp consists basically of two differential amplifiers, and a single-ended output circuit in cascade (the output of one stage feeding the input of another stage). The pair of cascaded differential amplifiers provides most of the gain.

Input circuits. The inputs to the op-amp are applied to the bases of the pair of emitter-coupled transistors Q_1 and Q_2 in the first differential amplifier. The inverting input is applied to the base of transistor Q_2, and the non-inverting input is applied to the base of Q_1. With a signal applied only to the inverting input and with the non-inverting input grounded, the output is an amplified and inverted version of the input, as discussed in Sec. 1-1.1.

First differential stage. Transistors Q_1 and Q_2 develop driving signals for the second differential amplifier. Transistor Q_6, a constant-current device, is included in the first differential stage to provide bias stabilization for Q_1 and Q_2. If there is an increase in supply voltage, which would normally increase the current through Q_1 and Q_2, the reverse-bias on Q_6 increases. This reduces the Q_1-Q_2 current and offsets the effect of the initial supply voltage increase.

Diode D_1 provides thermal compensation for the first differential stage. If there is an increase in operating temperature for the op-amp, which would normally increase the current through Q_1 and Q_2, there is a corresponding increase in current flow through D_1, since the diode is fabricated on the same silicon chip as the transistors. The increase in

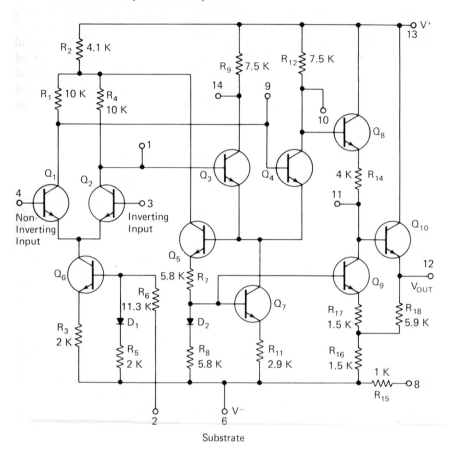

Fig. 1-5. RCA CA3008 op-amp.

D_1 current also increases the reverse-bias of Q_6, thus offsetting the change in temperature.

Second differential stage. The emitter-coupled transistors, Q_3 and Q_4, in the second differential amplifier are driven push-pull by the outputs from the first differential amplifier. Bias stabilization for the second differential amplifier is provided by constant-current transistor Q_7. Compensating diode D_2 provides the thermal stabilization for the second differential amplifier, and also for current-sink transistor Q_9 in the output stage.

Negative feedback stage. Transistor Q_5 develops a *negative feedback* to reduce common-mode error signals that are developed when the same input is applied to both input terminals of the op-amp. Transistor Q_5 samples the signal that is developed at the emitters of Q_3 and Q_4. Because the second differential stage is driven push-pull, the signal at this point will be zero when the first differential stage and the base-emitter circuits of the second stage are matched, and there is no common-mode input.

Common-mode signal rejection. A portion of any common-mode, or error, signal that appears at emitters of transistors Q_3 and Q_4 is developed by transistor Q_5 across resistor R_2 (the common collector resistor for transistors Q_1, Q_2 and Q_5) in the proper phase to reduce error. The emitter circuit of transistor Q_5 also reflects a portion of the same error signal into constant current transistor Q_7 in the second differential stage so that the initial error is further reduced.

Power supply variation rejection. Transistor Q_5 also develops feedback signals to compensate for common-mode effects produced by variations in the power supply. For example, a decrease in the positive power supply voltage results in a decrease of the voltage at the emitters of Q_3 and Q_4. This negative-going change in voltage is reflected by the emitter circuit of Q_5 to the bases of constant-current transistors Q_7 and Q_9. Less current then flows through these transistors. The decrease in collector current of Q_7 results in a reduction of the current through Q_3 and Q_4, and the collector voltages of these transistors tend to increase. This tendency partially cancels the decrease that occurs with the reduction of the positive supply voltage.

The partially cancelled decrease in the collector voltage of Q_4 is coupled directly to the base of Q_8 and is transmitted by the emitter of Q_8 to the base of Q_{10}. At this point, the decrease in voltage is further cancelled by the increase in collector voltage of current-sink transistor Q_9.

In a similar manner, transistor Q_5 develops the compensating feedback to cancel the effects of an increase in the positive supply (or of variation in the negative supply voltage). Because of the feedback stabilization provided by transistor Q_5, the op-amp of Fig. 1-5 provides high common-mode rejection, and has excellent open-loop stability and a low sensitivity to power supply variations. All of these characteristics are critical to op-amp operation, as is discussed in following sections of this chapter.

Output circuit. In addition to their function in the cancellation of supply voltage variations, transistors Q_8, Q_9 and Q_{10} are used in an emitter-follower type of single-ended output circuit. The output of the second differential amplifier is directly coupled to the base of Q_8. The emitter circuit of transistor Q_8 supplies the base-drive for output transistor Q_{10}.

A small amount of signal gain in the output circuit is made possible by the connection from the emitter of output transistor Q_{10} to the emitter circuit of transistor Q_9 (at the junction of R_{16} and R_{17}). If this connection were omitted, transistor Q_9 could be considered as merely a constant-current device for drive transistor Q_8. Because of the connection, however, the output circuit can provide a signal gain of 1.5 from the collector of differential amplifier transistor Q_4 to the output. Although this small amount of gain may seem insignificant, it does increase the output-swing capabilities of the op-amp.

The output from the op-amp circuit is taken from the emitter of the output transistor Q_{10} so that the dc level of the output signal is substantially lower than that of the differential amplifier output at the collector of transistor Q_4. In this way, the output circuit shifts the dc level at the output so that it is effectively the same as that at the input when no signal is applied. This problem of dc level shifting is discussed further in Sec. 1-1.5.

Resistor R_{15} in series with terminal 8 increases the ac short-circuit load capability of the op-amp when terminal 8 is shorted to terminal 12 so that R_{15} is connected between the output and the negative supply.

1-1.5 DC level shifting problems in op-amps

In any cascade direct-coupled amplifier, either discrete component or IC, the dc level rises through successive stages toward the supply voltage. In linear IC op-amps, the dc voltage builds up through the NPN stages in the positive direction and must be shifted negatively if large output signal swings are to be obtained. For example, if the supply voltage is 10V and the output is at 9V under no-signal conditions, the maximum output voltage swing is limited to less than 1V.

Because op-amps use external feedback, it is especially important to provide for compensation of the dc level shift. Op-amps must have equal (and preferably zero) input and output dc levels so that the dc coupling of the feedback connection does not shift any bias point. For example, if the input is at 0V, but the output is at 3V, the 3V is reflected back through the external feedback resistor and changes the input to a 3V level.

The use of an output stage, such as shown in Fig. 1-5, is a common technique to prevent a shift in dc level between the output and input of an op-amp. Transistor Q_8 operates as an input buffer, and transistor Q_9 is essentially a current sink for Q_8. The shift in dc level is accomplished by the voltage drop across resistor R_{14} produced by the collector current of transistor Q_9. The emitter of the output transistor Q_{10} is connected to the emitter of Q_9 (through R_{18} at the junction of R_{17} and R_{16}). Feedback through R_{18} results in a decrease in the voltage drop across R_{14} for negative-going output swings, and an increase in this voltage drop for positive-going output swings.

If properly designed, the circuit shown in Fig. 1-5 can provide substantial voltage gain, high input impedance, low output impedance and an output swing *nearly equal* to the supply voltages, in addition to the desired shift in dc level. Moreover, feedback may be coupled from the output to the input to compensate for dc common-mode effects that result from variations in the supply voltages.

1-1.6 High-performance op-amps

The circuit of Fig. 1-6 provides improvements over the circuit of Fig. 1-5. The op-amp shown in Fig. 1-6 consists of two differential amplifier stages followed by a class B output stage. Emitter-follower inputs provide an exceptionally high input impedance (typically 1.5 megohms), and low bias current. (Terms such as input impedance and bias current are defined in Sec. 1-3.) An additional advantage of the emitter-follower inputs is that the Miller capacitance of the differential amplifier is substantially reduced, and the input capacitance of the op-amp is lower than if a similar single-transistor configuration were used.

The output of the first differential amplifier stage (Q_3 and Q_4) is buffered from the input of the next differential amplifier stage (Q_6 and Q_7) by emitter followers Q_{19} and Q_{20}. This arrangement reduces the input loading effects on the first stage and therefore maintains the first stage gain.

A circuit is incorporated in this design to sense any change in the operating point of the first differential amplifier caused by variations in either the positive or the negative supply voltage. Any changes in the supply voltages are reflected to the base of transistor Q_5, which detects changes in the collector voltage of the first differential amplifier and compensates for them.

For example, a rise in the voltage at the emitters of Q_6 and Q_7 increases the bias voltage to Q_5, and thus increases the collector current, countering the apparent rise in the collector voltage of either Q_3 and Q_4. At the same time, the emitter current of Q_5 also increases, and increases the voltage drop across the diode-connected transistor Q_9 and resistor R_{10} to increase the collector current of Q_8. Thus, any apparent increase in the collector voltage of the first differential stage causes a correction both at the constant-current source Q_8 and at the collector supply voltage through R_1, the common load resistor for Q_3, Q_4 and Q_5.

An emitter-follower Q_{11} buffers the output of the second differential amplifier stage, and drives the divider and summing network to the output stage. Resistor R_{13} may be considered the input resistance of an amplifier to the summing point, the junction of R_{13}, R_{14} and R_{15}. Resistor R_{14} shifts to the operating point of the output stage with little attenuation of the signal as a result of the high collector impedance of the constant-current source Q_{12}.

Diode-connected transistors Q_{13A} and Q_{13} provide further dc shifting

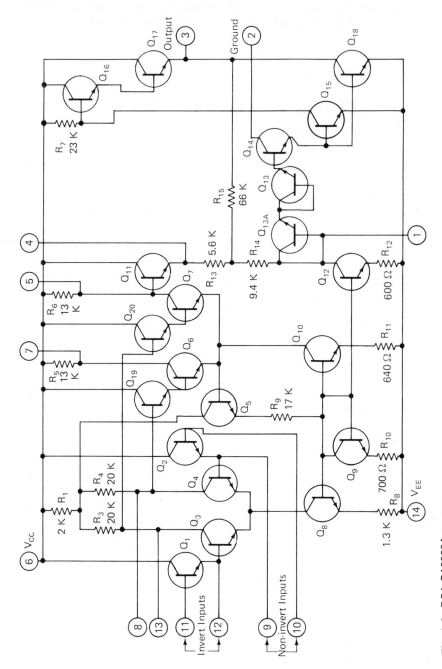

Fig. 1-6. RCA CA3033A op-amp.

of the signal to the base of emitter-follower Q_{14}. This emitter-follower provides further level shifting and a low driving impedance to transistors Q_{15} and Q_{18}. The input to the base of Q_{13A} from terminal 1 provides a means of controlling the op-amp from an external source. The op-amp is designed for systems in which the circuit must be disabled temporarily, say by a "squelch" or disable pulse. For example, a pulse equal to V_{CC} will completely disable the op-amp.

The excellent matching of the base-to-emitter voltage (that is typical of IC transistors) makes it possible to establish the idling current of the output stage accurately. Because the collector-current characteristics of Q_{15} and Q_{18} (as a function of base-to-emitter voltage) are matched, the collector current in Q_{15} determines the idling current of Q_{18}.

For example, if the operating current of Q_{15} is set at 1 mA for a given base-to-emitter voltage, the operating current of Q_{18} is also 1 mA because the base-to-emitter voltages of both transistors are the same. This type of design is not usually possible in discrete component op-amps since it is difficult to get the exact base-to-emitter voltage match (except in IC form on a monolithic chip).

1-1.7 Variable-bias (micropower) op-amps

Variable-bias op-amps are similar to conventional op-amps, except that standby power can be controlled by an external bias voltage. The bias input is generally introduced at the constant-current transistor for the input differential amplifier. This is shown in Fig. 1-7, which is the schematic for a variable-bias op-amp in IC form. The bias sets the quiescent current (no-signal current, standby current, operating current, etc.) of the entire op-amp. Thus, the operating point of the op-amp can be controlled by external means.

An increase in bias current produces an increase in quiescent current. For example, as shown in Fig. 1-8, a variation from 1 to 400 nA of input bias (shown as I_{IB}) produces a corresponding variation in quiescent current (I_Q) of 1 to 800 μA. Thus, an IC op-amp with variable bias can deliver milliamperes of current, yet only consume microwatts of standby power. For this reason, some variable-bias op-amps are known as *micropower* op-amps.

The variation in I_Q produces changes in other op-amp characteristics (which are defined in Sec. 1-3). For example, as shown in Fig. 1-9, input offset current I_{IO} increases directly with I_Q. However, input offset voltage is not greatly affected by I_Q, as shown in Fig. 1-10. An increase in I_Q will produce input offset voltage slightly.

The maximum output capability of the op-amp is affected by I_Q, but

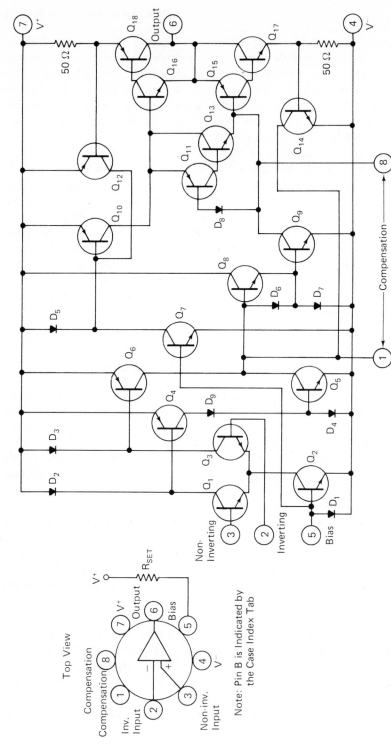

Fig. 1-7. RCA CA3078T micropower op-amp.

16

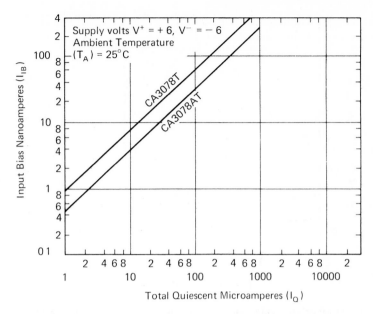

Fig. 1-8. Input bias current versus total quiescent current (Courtesy RCA).

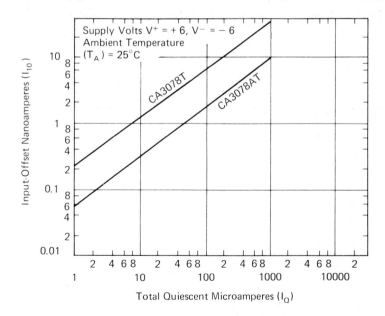

Fig. 1-9. Input offset current versus total quiescent current (Courtesy RCA).

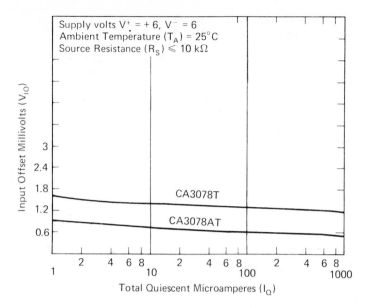

Fig. 1-10. Input offset voltage versus total quiescent current (Courtesy RCA).

not to the extent of input offset current. As shown in Fig. 1-11, maximum output current capability increases up to a point for an increase in I_Q. Once that point is reached (at about 6 μA) further increases in I_Q produce no further increases in maximum output current capability.

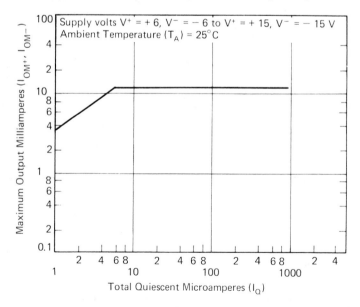

Fig. 1-11. Maximum output current versus total quiescent current (Courtesy RCA).

The I_Q has little effect on voltage gain of the op-amp, as shown in Fig. 1-12. This is especially true when the load resistance is large. Load resistance is also the dominating factor in output voltage swing, as shown in Fig. 1-13.

The bias used to control the op-amp can be obtained from any source. Generally, the bias is taken from the $V+$ supply through a fixed resistance. Figure 1-14 shows how this is done (for the op-amp of Fig. 1-7) using an external resistor R_{SET}. Figure 1-14 also shows a graph used to determine the value of the bias resistance R_{SET}.

Because of the low power consumption possible with variable-bias op-amps, battery operation is possible. Figure 1-15 shows an IC variable-bias op-amp used as a 20 dB amplifier powered from a 1.5V AA battery cell.

1-1.8 Summary of advantages for differential op-amps

The balanced differential amplifier is considered the best configuration for general-purpose IC op-amps by most manufacturers for the reasons outlined thus far. The following are some additional reasons for use of differential amplifiers in op-amps:

The differential amplifier allows for injection of an adjustable dc voltage into either channel. This provides for compensation of unbalance in the op-amp, and permits both input and output levels to be set at zero.

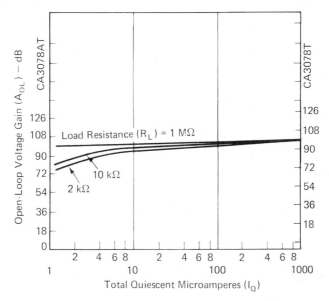

Fig. 1-12. Open-loop voltage gain versus total quiescent current (Courtesy RCA).

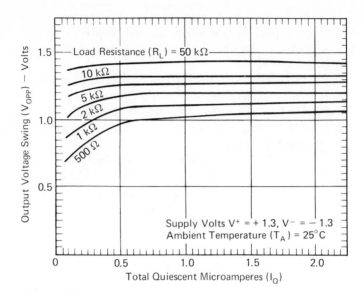

Fig. 1-13. Output voltage swing versus total quiescent current (Courtesy RCA).

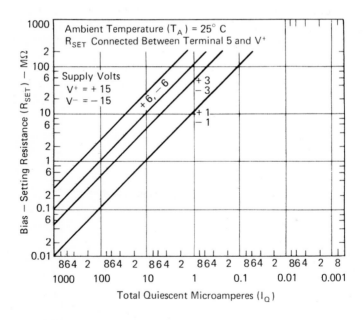

Fig. 1-14. Bias setting resistance versus total quiescent current (Courtesy RCA).

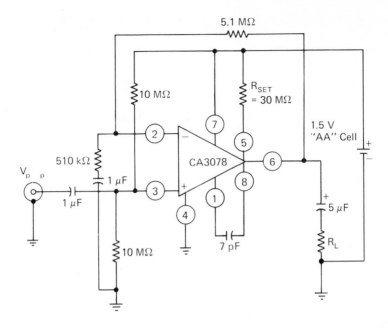

Fig. 1-15. Non-inverting 20 dB amplifier using micropower op-amp (Courtesy RCA).

The same provision permits establishment of a new voltage reference level at the output (at some point other than zero) if desired.

Advantage can be taken from the exceptional balance between the differential inputs that results from the inherent match in base-to-emitter voltage and short-circuit current gain of the two (differential-pair) transistors that are processed in exactly the same way, and are located very close to each other on the same small silicon chip.

The differential amplifier uses a minimum number of capacitors. Generally, no internal capacitors are used in IC op-amps. An exception to this is where one internal capacitor is used for phase compensation, as discussed in Sec. 1-2.

The use of large resistors can usually be avoided, and the gain of the differential amplifier circuit is a function of resistance ratios, rather than of actual resistance values.

The differential amplifier is much more versatile than other possible circuit configurations and can be readily adapted for use in a variety of component applications. For example, in the circuits of Figs. 1-5, 1-6 and 1-7, there are many *connections to internal circuit components,* in addition to input, output and power supplies. This is typical of differential amplifier IC op-amps. The additional connections (such as the collectors of Q_1, Q_2, Q_3, Q_4 and Q_9 in Fig. 1-5) provide a variety of input-output points for the phase compensation techniques described in Sec. 1-2.

1-2. FREQUENCY RESPONSE (BANDWIDTH) AND GAIN

Most of the design problems for op-amps are the result of tradeoffs between gain and frequency response (or bandwidth). The *open-loop* (without feedback) gain and frequency response are characteristics of the basic op-amp circuit, but they can be modified with *phase compensation* networks. The *closed-loop* (with feedback) gain and frequency response are primarily dependent upon *external feedback* components. Although op-amps are generally used in the closed-loop operating mode, the open-loop characteristics have considerable effect on

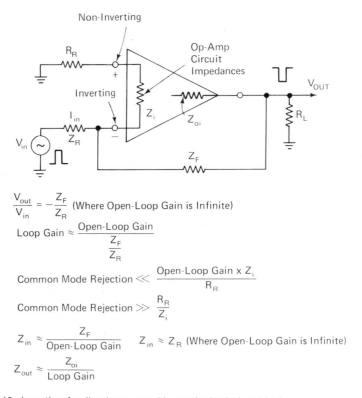

$$\frac{V_{out}}{V_{in}} = -\frac{Z_F}{Z_R} \text{ (Where Open-Loop Gain is Infinite)}$$

$$\text{Loop Gain} \approx \frac{\text{Open-Loop Gain}}{\dfrac{Z_F}{Z_R}}$$

$$\text{Common Mode Rejection} \ll \frac{\text{Open-Loop Gain} \times Z_i}{R_R}$$

$$\text{Common Mode Rejection} \gg \frac{R_R}{Z_i}$$

$$Z_{in} \approx \frac{Z_F}{\text{Open-Loop Gain}} \qquad Z_{in} \approx Z_R \text{ (Where Open-Loop Gain is Infinite)}$$

$$Z_{out} \approx \frac{Z_{oi}}{\text{Loop Gain}}$$

Fig. 1-16. Inverting feedback op-amp (theoretical relationships).

operation, and must be considered when designing a feedback system for an op-amp.

1-2.1 Inverting and noninverting feedback

The two basic op-amp feedback systems, inverting feedback and noninverting feedback, are shown in Figs. 1-16 and 1-17, respectively. In both cases, the op-amp output is fed back to the inverting (or minus) input through an impedance Z_F to control frequency response and gain. In the inverting feedback system of Fig. 1-16, the input signal is also applied to the inverting input, resulting in an inverted output. In the noninverting system of Fig. 1-17, the input signal 's applied to the noninverting input, producing a noninverted output (a positive input produces a positive output). The inverting feedback system of Fig. 1-16, in which a positive input produces a negative output, is the more commonly used.

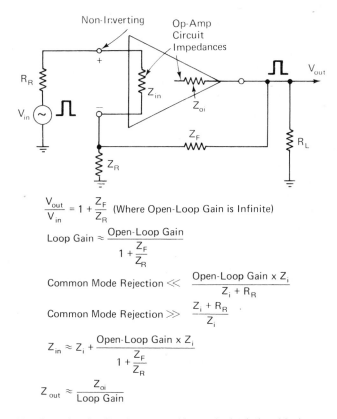

$$\frac{V_{out}}{V_{in}} = 1 + \frac{Z_F}{Z_R} \quad \text{(Where Open-Loop Gain is Infinite)}$$

$$\text{Loop Gain} \approx \frac{\text{Open-Loop Gain}}{1 + \frac{Z_F}{Z_R}}$$

$$\text{Common Mode Rejection} \ll \frac{\text{Open-Loop Gain} \times Z_i}{Z_i + R_R}$$

$$\text{Common Mode Rejection} \gg \frac{Z_i + R_R}{Z_i}$$

$$Z_{in} \approx Z_i + \frac{\text{Open-Loop Gain} \times Z_i}{1 + \frac{Z_F}{Z_R}}$$

$$Z_{out} \approx \frac{Z_{oi}}{\text{Loop Gain}}$$

Fig. 1-17. Non-inverting feedback op-amp (theoretical relationships).

The equations shown in Figs. 1-16 and 1-17 are classic guidelines. The equations do not take into account the fact that open-loop gain is not infinitely high and output impedance is not infinitely low. Thus, the equations contain built-in inaccuracies and must be used as *guides only*.

With both feedback configurations, the closed-loop gain, or the ratio of V_{OUT} to V_{IN}, is *approximately* equal to the ratio of Z_F to Z_R. Typically, Z_F and Z_R are fixed, composition (noninductive) resistors. Thus, if Z_F is a 100,000 ohm resistor, and Z_R is 1000 ohms, the resistance ratio is 100 to 1, and the approximate closed-loop gain is 100 (40 dB).

In theory, any voltage gain could be set by proper selection of Z_F and Z_R. There are obvious practical limits. For example, if the open-loop gain is less than 100, there is no ratio of Z_F and Z_R that will produce a gain of 100. Likewise, even if the gain can be obtained, an improper ratio of Z_F and Z_R could operate the op-amp in an unstable condition (as is discussed in later paragraphs of this section).

With both configurations, *loop gain* is defined as the ratio of open-loop gain to closed-loop gain. When loop gain is large, the inaccuracies in the equations will decrease. For example, if open-loop gain is 1000 and closed-loop gain is 100, loop gain is 10. If closed-loop gain is reduced to 10, loop gain is increased to 100, and the equations of Figs. 1-16 and 1-17 are more accurate. However, as is discussed in later paragraphs, it is not always possible, nor is it always desirable, to operate with a large loop gain.

In the inverting configuration of Fig. 1-16, the input impedance of the system Z_{IN} (that appears to the input signal) is approximately equal to resistance Z_R. The output impedance of the system Z_{OUT} is approximately equal to the intrinsic input impedance of the op-amp Z_i, divided by the loop gain. The common mode rejection is some value much greater than the ratio of R_R and Z_i. Thus, a large value of R_R is desirable for good common-mode rejection. However, it is not always possible to use large values of R_R.

In the non-inverting configuration of Fig. 1-17, the system characteristics are similar to those for inverting feedback, as shown by the equations. The major difference is that the op-amp intrinsic input impedance Z_i has a greater effect in the case of noninverting (Fig. 1-17) operation.

1-2.2 Gain/frequency characteristics of a theoretical op-amp

The gain/frequency relationships shown in Fig. 1-18 are based on a theoretical op-amp. That is, the open-loop gain remains flat as frequency increases, and then begins to roll off at some particular

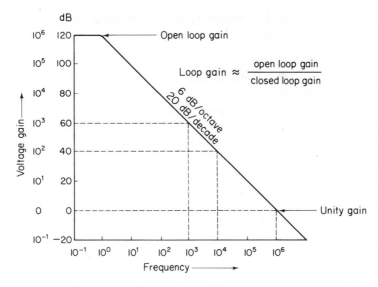

Fig. 1-18. Frequency response curve of theoretical op-amp.

frequency. The point at which the rolloff starts is sometimes referred to as a *pole*. The curve of Fig. 1-18 is known as a *one pole* plot (and may also be known as a Bode plot, gain/bandwidth plot or a frequency response curve). At the first (and only) pole, the open-loop gain rolls off at 6 dB per octave, or 20 dB per decade. The term 6 dB/octave means that the gain drops by 6 dB each time frequency is doubled. This is the same as a 20-dB drop each time the frequency is increased by a factor of 10.

If the open-loop gain of an amplifier is as shown in Fig. 1-18, any stable closed-loop gain could be produced by the proper selection of feedback components, provided the closed-loop gain is less than the open-loop gain. The only concern would be a tradeoff between gain and frequency response.

For example, if a gain of 40 dB (10^2) is desired, a feedback resistance Z_F that is 10^2 times larger than the input resistance Z_R is used (such as a Z_F of 1000 and a Z_R of 10). The closed-loop gain is then flat to 10^4 Hz, and rolls off at 6 dB/octave to unity gain (a gain of 1) at 10^6 Hz. If a 60 dB (10^3) gain is required instead, the feedback resistance Z_F is raised to 10^3 times the input resistance Z_R (Z_F of 10,000 and a Z_R of 10). This reduces the frequency response. Gain is flat to 10^3 Hz (instead of 10^4), followed by rolloff of 6 dB/octave down to unity gain.

It should be noted that the gain/frequency response curves of practical op-amps rarely look like that shown in Fig. 1-18. A possible exception is an internally compensated IC op-amp, such as described in later paragraphs.

1-2.3 Gain/frequency characteristics of a practical op-amp

The open-loop frequency response curve of a practical op-amp more closely resembles that shown in Fig. 1-19. This curve is a three pole plot. The first pole occurs at about 0.2 MHz, the second pole at 2 MHz, and the third pole at 20 MHz. (In a truly practical frequency response curve there will be no sharp breaks at the poles. Instead, the pole "corners" will be rounded and often difficult to distinguish. However, the curve of Fig. 1-19 is given here to illustrate certain frequency response characteristics.)

In the curve of Fig. 1-19 gain is flat at 60 dB to the first pole, then rolls off to 40 dB at the second pole. Since there is a decade between the first and second poles (0.2 MHz to 2 MHz), the rolloff is 20 dB per decade, or 6 dB/octave. As frequency increases, rolloff continues from the second pole to the third pole, where gain drops from 40 dB to 0 dB. Thus, the rolloff is 40 dB/decade or 12 dB/octave. At the third pole, gain drops below unity as frequency increase at a rate of 60 dB/decade or 18 dB/octave.

Some op-amp datasheets provide a curve similar to that shown in Fig. 1-19. If the datasheets are not available, it is possible to test the op-amp under laboratory conditions, and draw an actual response curve (frequency response and phase shift). The necessary procedures are described in Chapter 5.

1-2.4 Phase shift problems

Note that the curve of Fig. 1-19 also shows phase shift of an op-amp. As frequency increases, the phase shift between input and output signals of the op-amp increases. At frequencies up to about 0.02 MHz, the phase shift is zero. That is, the output signals are in phase with the noninverting input, and exactly 180° out of phase with the inverting input. As frequency increases up to about 0.2 MHz, phase shift increases by about 45°. That is, the output signals are 45° out of phase from the noninverting input, and 45° from 180° (or 135°) out of phase with the inverting input. At about 6 MHz, the output signals are 180° out of phase with the noninverting input, and in phase with the inverting input. Since the output is fed back to the inverting input through Z_F, the input and output are in phase. If the output amplitude is large enough, the op-amp will then oscillate.

Oscillation can occur if the output is shifted near to 180° (and fed back to the inverting input). Even if oscillation does not occur, op-amp

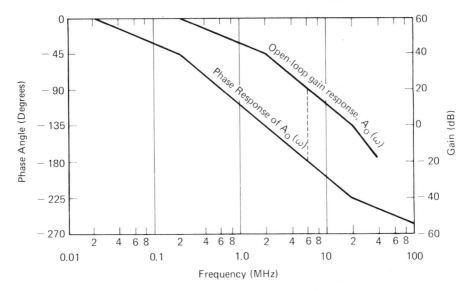

Fig. 1-19. Gain and phase response of an open-loop op-amp without phase compensation (Courtesy RCA).

operation can become unstable. For example, the gain will not be flat. Usually a *peaking condition* will occur, in which output remains flat up to a frequency near the 180° phase shift point, and then gain will increase sharply to a peak. This is caused by the output signals being nearly in phase with input signals and reinforcing the input signals. At higher frequencies, gain will drop off sharply and/or oscillation will occur. *As a guideline, the op-amp should never be operated at a frequency at which phase shift is near 180° (above 170°) without compensation.* In the op-amp of Fig. 1-19, the maximum uncompensated frequency would be about 5 MHz. This is the frequency at which open-loop gain is about +25 dB.

One guideline often mentioned in op-amp literature is based on the fact that the 180° phase shift point almost always occurs at a frequency at which the open-loop gain is in the 12 dB/octave slope. Thus, the guideline states that a closed-loop gain should be selected so that the unity gain is obtained at some frequency near the beginning of the 12 dB/octave slope (such as at 2 MHz in the op-amp of Fig. 1-19). However, this usually results in a very narrow bandwidth.

A more practical guide can be stated as follows: when a selected closed-loop gain is equal to or less than the open-loop gain at the 180° phase-shift point, the op-amp will be unstable. For example, if a closed-loop gain of 20 dB or less is selected, an op-amp with open-loop, uncompensated curves similar to Fig. 1-19 will be unstable.

To find the minimum closed-loop gain, simply note where the $-180°$ phase angle intersects the phase shift line. Then draw a vertical line up to cross the open-loop gain line. The closed-loop gain must be more than the open-loop gain at the frequency where the $180°$ phase shift occurs, but less than the maximum open-loop gain. Using Fig. 1-19 as an example, the closed-loop gain would have to be greater than 20 dB, but less than 60 dB.

Keep in mind that the guidelines discussed thus far apply to an uncompensated op-amp. With proper phase compensation, bandwidth (frequency response) and/or gain can be extended.

1-2.5 Op-amp phase compensation methods

Op-amp design problems created by excessive phase shift can be solved by compensating techniques that alter response so that excessive phase shifts no longer occur in the desired frequency range. The following are the basic methods of phase compensation.

Closed-loop modification. The closed-loop gain of an op-amp can be altered by means of capacitors and/or inductances in the external feedback circuit (in place of fixed resistances). Capacitors and inductances change impedance with changes in frequency. This provides a different amount of feedback at different frequencies, and changes the amount of phase shift in the feedback signal. The capacitors and inductances can be arranged to offset the undesired open-loop phase shift.

Phase shift compensation by closed-loop modification is generally not recommended since the method can create impedance problems at both the high and low frequency limits of operation. However, closed-loop modification is used for applications in which the op-amp is part of a bandpass, band rejection or peaking filter, as are described in Chapter 2.

Input impedance modification. The open-loop input impedance of an op-amp can be altered by means of resistors and capacitors connected at the op-amp input terminals. The impedance presented by the RC combination changes with frequency, thus altering the input impedance of the op-amp. In turn, the change in input impedance (with frequency) changes the bandwidth and phase shift characteristics of the op-amp. Such an arrangement causes the rolloff to start at a lower frequency than the normal open-loop response of the op-amp, but produces a *stable rolloff* similar to that of the "ideal" curve in Fig. 1-18. With an op-amp properly compensated by an RC circuit at the input, the desired closed-loop gain can be produced by selection of external feedback resistors in the normal manner.

Phase compensation techniques that alter the open-loop input impedance permit the introduction of a zero into the response. This zero can be designed to cancel one of the poles in the open-loop response. Typically, the first pole is cancelled, and the open-loop gain drops to zero at the second pole. That is, after modification, the response drops to zero at a frequency where the uncompensated response changes from 6 dB/octave to 12 dB/octave. This is shown in Fig. 1-20. In another, less-frequently used input modification design, the response drops to zero at the frequency where the uncompensated response changes from 12 dB/octave to 18 dB/octave. Both input impedance modification designs are discussed in Sec. 1-2.6.

Phase-lead compensation. The open-loop gain and phase shift characteristics of an op-amp can be modified by means of a capacitor (or capacitors) connected to stages in the op-amp. Usually, the capacitors are connected between collectors in one of the high-gain differential stages. In other cases, the capacitors are connected from the collectors to ground. Generally, the capacitors are external to the op-amp, and are connected to the internal stages by means of terminals provided on the package (such as the terminals shown in Figs. 1-5 through 1-7).

Phase-lead compensation requires a knowledge of the op-amp circuit characteristics. Usually, information for phase-lead compensation is provided on the op-amp datasheet. Typical phase-lead systems are described in Sec. 1-2.7.

Phase-lag compensation. The open-loop gain and phase shift characteristics of an op-amp can be modified by means of a series capacitor and resistor connected to stages in the op-amp. There are two basic phase-lag compensation systems.

In one system, generally known as RC *rolloff, straight rolloff* or *phase-lag rolloff compensation,* the open-loop response is altered by means of an RC network connected across a circuit component, such as across the input or output of an op-amp gain stage.

In the other system, generally known as *Miller-effect rolloff* or *Miller-effect phase-lag compensation,* the open-loop response is altered by means of an RC network connected between input and output of an inverting gain stage in the op-amp. The impedance of the compensating RC network then appears to be divided by the gain of that stage.

With either method, the rolloff starts at the corner frequency produced by the RC network. The Miller-effect rolloff technique requires a much smaller phase-compensating capacitor than that which must be used with the straight rolloff method. Also, the reduction in swing capability that is inherent in the straight rolloff is delayed significantly when the Miller-effect rolloff is used.

As with phase-lead, either method of phase-lag compensation requires a knowledge of the op-amp circuit characteristics. Usually, information for phase-lag compensation is provided on the op-amp datasheet, when such methods are recommended by the manufacturer. A typical straight rolloff system is described in Sec. 1-2.7. Miller-effect rolloff is discussed in Sec. 1-2.8.

How to select a phase compensation method. A comprehensive op-amp datasheet will recommend one or more methods for phase compensation and will show the relative merits of each method. Usually this is done by means of response curves for various values of the compensating network. Several examples are described in Sec. 1-2.6 through 1-2.8.

The *recommended* phase compensation methods and values should be used in all cases. Proper phase compensation of an op-amp is at best a difficult, trial-and-error job. By using the datasheet values it is possible to take advantage of the manufacturers' test results on production quantities of a given op-amp.

If the datasheet is not available or if the datasheet does not show the desired information, it is still possible to design a phase compensating network using rule-of-thumb equations.

The first step in phase compensation (when not following the datasheet) is to test the op-amp for open-loop frequency response and phase shift, as described in Chapter 5. Then draw a response curve similar to that in Fig. 1-19. On the basis of actual open-loop response, and the information in Secs. 1-2.5 through 1-2.8, select trial values for the phase compensating network. Then repeat the frequency response and phase shift tests. If the response is not as desired, change the values as necessary.

Each method of phase compensation has its advantages and disadvantages. The main advantage of open-loop input impedance modification (Sec. 1-2.6) is that it can be accomplished without datasheet information (or with limited information). The only op-amp characteristic required is input impedance. This is almost always available in datasheet or catalog form. If not, input impedance can be found by a simple test described in Chapter 5.

Phase-lead and phase-lag are the most widely accepted techniques for op-amps (particularly IC op-amps). These methods have an advantage (over input modification) in that the phase compensation network is completely isolated from the feedback network. In the case of input modification, resistance in the phase compensation network forms part of feedback network.

Phase-lead and phase-lag compensation have certain disadvantages. A careful inspection of the information in Secs. 1-2.7 through 1-2.9 will show that it is necessary to know certain internal characteristics of the op-amp before an accurate prediction of the compensated frequency response can be found. In the case of phase-lead (Sec. 1-2.7), the compensated response is entirely dependent upon value of the capacitors, and must be found by actual test. In the case of phase-lag compensation (Secs. 1-2.8 and 1-2.9), the values for R and C of the compensation network are based on the uncompensated open-loop frequency at which gain changes from a 6 dB/octave drop to a 12 dB/octave drop. This can be found by test of the uncompensated op-amp. However, to predict the frequency at which the compensated response will start to roll off (or the gain after compensation) requires a knowledge of internal-stage transconductance (or gain) and stage load. This information is usually not available and can not be found by simple test.

To sum up, if the datasheet is available, use the recommended phase compensation method. It will probably be phase-lead or phase-lag. If no phase compensation information is available, use input impedance modification.

1-2.6 Phase compensation by modification of input impedance

There are two accepted methods for phase compensation using input impedance modification. The first method, shown in Fig. 1-20, is the most widely used since it provides a straight rolloff similar to the ideal curve of Fig. 1-18. Once the input circuit is modified, conventional feedback can be used to select any point along the rolloff. That is, any combination of gain and frequency can be produced as described in Sec. 1-2.2. The main disadvantage to the method of Fig. 1-20 is early rolloff. Thus, if high gain is required, the bandwidth will be very narrow.

The method shown in Fig. 1-21 is used only where bandwidth is of greatest importance, and gain can be sacrificed. As shown, rolloff does not *start* until the breakpoint between 6 dB/octave and 12 dB/octave is reached, and gain is flat up to that point (no peaking condition). However, the method of Fig. 1-21 usually results in little gain across the operating frequency range.

Early rolloff method. Assume that the method of Fig. 1-20 is to be used with an op-amp having the characteristics shown in Fig. 1-19. That is, the uncompensated, open-loop gain is 60 dB, the 6 dB/octave rolloff starts at about 0.2 MHz, the 12 dB/octave starts at 2 MHz and the 18 dB/octave starts at 20 MHz (which is also the point at which the open-loop gain drops to zero).

The first step in using the method shown in Fig. 1-20 is to note the frequency at which the uncompensated rolloff changes from 6 dB to 12 dB (point A of Fig. 1-20). The compensated rolloff should be zero (unity gain, point B) at the same frequency.

Draw a line up to the left from point B that *increases* at 6 dB/octave.

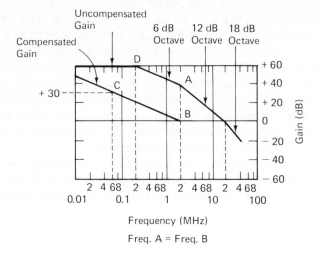

Freq. A = Freq. B

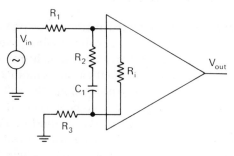

R_i = Input Impedance of Op-Amp

$R_1 = R_3$

$R_1 + R_3 = \left(\dfrac{\text{Uncompensated Gain (dB)}}{\text{Compensated Gain dB}} \right) R_i$

$R_2 = \dfrac{R_1 + R_3}{\left(\dfrac{\text{Freq. D}}{\text{Freq. C}} - 1 \right) \left(1 + \dfrac{R_1 + R_3}{R_i} \right)}$

$C_1 = \dfrac{1}{6.28 \times \text{Freq. D} \times R_2}$

Compensated Gain $= \dfrac{\text{Uncompensated Gain} \times R_i}{R_i + R_1 + R_3}$

Freq. D $= \dfrac{1}{6.28 \times R_2 \times C_1}$

Fig. 1-20. Phase compensation by modification of input impedance (early rolloff method).

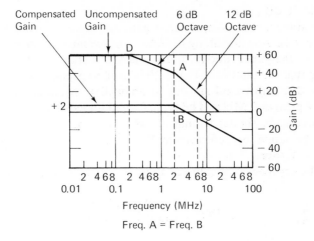

Freq. A = Freq. B

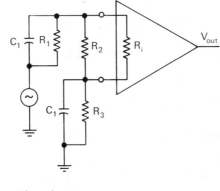

R_i = Input Impedance
   of Op-Amp

$$R_1 = R_3 = \frac{\left(\dfrac{1}{6.28 \times \text{Freq. D}}\right)}{C_1}$$

$$R_2 = \frac{\text{Compensated Gain} \times 2R_1}{\text{Uncompensated Gain}}$$

$$\text{Fréq. C} \approx \frac{\left(\dfrac{2}{C_1 \, R_2}\right)}{6.28}$$

C_1 = See Text

Fig. 1-21. Phase compensation by modification of input impedance (extended band-width method).

For example, with point B at 2 MHz, the line should intersect 0.2 MHz as it crosses the 20 dB gain point, and should intersect 0.02 MHz as it crosses the 40 dB point.

Any combination of compensated gain and rolloff starting frequency (point C) can be selected along the line. For example, if the rolloff starts at 0.2 MHz, the gain is about 20 dB, and vice versa.

Assume that the circuit of Fig. 1-20 is used to produce a compensated gain of 30 dB, with rolloff starting at about 0.06 MHz and dropping to zero (unity gain) at 2 MHz. The typical input impedance R_i is 10K. (Uncompensated gain, similar to that of Fig. 1-19, and typical input impedance can be found by referring to the datasheet, or by actual test as described in Chapter 5.)

Using the compensated gain equation of Fig. 1-20, the relationship is:

$$30 \text{ dB} = \frac{(60 \text{ dB}) \ (10{,}000)}{(10{,}000 + R_1 + R_3)}$$

Therefore,

$$R_1 + R_3 = \left(\frac{60}{30} - 1\right) 10{,}000$$

$$= 1 \times 10{,}000$$

$$R_1 + R_3 = 10{,}000$$

$$\text{If } R_1 = R_3, \ R_1 = R_3 = 5000$$

Using the equation in Fig. 1-20, the value of R_2 is:

$$R_2 = \frac{10{,}000}{\left(\dfrac{0.2}{0.06} - 1\right)\left(1 + \dfrac{10{,}000}{10{,}000}\right)}$$

$$= \frac{10{,}000}{2.3 \times 2}$$

$$= 2100\Omega \text{ (nearest standard value)}$$

The value of C_1 is:

$$C_1 = \frac{1}{(2100) \ (6.28) \ (0.2 \text{ MHz})} \approx 0.0004 \ \mu\text{F (nearest standard value)}$$

If the circuit of Fig. 1-20 shows any instability in the open-loop or closed-loop condition, try increasing the values of R_1 and R_3 (to reduce gain); then select new values for R_2 and C_1.

Extended bandwidth method. Assume that the method of Fig. 1-21 is to be used with an op-amp having the characteristics shown in Fig. 1-19.

The first step in using the method shown in Fig. 1-21 is to note the frequency at which the uncompensated rolloff changes from 6 dB to 12 dB (point A of Fig. 1-21, 2 MHz). The compensated rolloff (point B) should start at the same frequency.

Assume that the circuit of Fig. 1-21 is used to produce a flat, compensated gain of 2 dB, with rolloff starting at 2 MHz. Find the approximate values for R_1 through R_3, and C_1. Also find the approximate maximum operating frequency (point C).

Assume a convenient value for each capacitor C_1, say 0.001 μF. Using this value, and the frequency (Freq. D) at which the uncompensated 6 dB rolloff starts (0.2 MHz), find the value of R_1 and R_3 as follows:

$$R_1 = R_3 = \frac{\left(\dfrac{1}{6.28 \times 0.2^{+6}}\right)}{0.001^{-6}}$$

$$\approx \frac{0.8^{-6}}{0.001^{-6}} \approx 800 \text{ ohms}$$

Using a desired compensated gain of 2 dB, an uncompensated gain of 60 dB, and a value of 1600 for $2R_1$, the value of R_2 is:

$$R_2 = \frac{2 \times 1600}{60} \approx 53 \text{ ohms}$$

Using 0.001 μF for C_1 and 53 ohms for R_2, the approximate maximum operating frequency (point C) is:

$$\text{Freq. C} = \frac{\left(\dfrac{2}{0.001^{-6} \times 53}\right)}{6.28} \approx 6 \text{ MHz}$$

1-2.7 Phase-lead compensation examples

As discussed, phase-lead compensation requires the addition of a capacitor (or capacitors) to the basic op-amp circuit. Generally the capacitors are external. However, some IC op-amps include an internal capacitor to provide fixed phase-lead compensation.

Phase-lead compensation requires a knowledge of internal op-amp circuit characteristics. As a result, the manufacturer's datasheet or some similar information must be used. The alternate method is to use typical values for the compensation capacitor, and test the results. Of course, this is time consuming, and may not prove satisfactory, even after tedious testing.

The following are some examples of manufacturers' data on phase-lead compensation.

Figure 1-22 shows typical phase-lead compensation characteristics for the op-amp of Fig. 1-6. The two compensating capacitors C_X and C_Y are connected from the collectors of the first differential amplifier (terminals 8 and 13, Fig. 1-6) to ground. When capacitance values greater than 0.1 μF are used, a lower-voltage capacitor that has a value equal to half that given on the curves may be connected between terminals 8 and 13, and a 0.001 μF capacitor connected from either terminal 8 or 13 to ground. This arrangement provides the same gain-phase rolloff shown on the curves of Fig. 1-22, and permits use of lower-voltage ceramic capacitors. For linear operation, the maximum expected difference voltage between the two collectors is less than 1V.

The dashed lines in Fig. 1-22 illustrate the use of the curves for design of a 60 dB amplifier. First, the intersection of the various gain-frequency

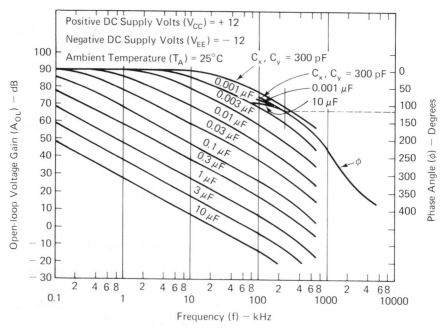

Fig. 1-22. Phase compensation characteristics for CA3033 or CA3033A (Courtesy RCA).

curves is followed out along the 60 dB line to the curve for a capacitor value of 0.001 μF. The intersection occurs at approximately 230 kHz. This means that if a 0.001 μF phase-lead capacitor is used, the op-amp response should be flat at 60 dB (within a 3 dB range) up to a frequency of 230 kHz. At higher frequencies, the op-amp output will drop off. Thus, any gain up to 60 dB can be selected by proper choice of feedback and input resistances.

Next, follow the 230 kHz line until it intersects the phase curve. The intersection occurs at approximately 118°. This means that if a 0.001 μF capacitor is used, and the op-amp is operated at 230 kHz, the phase shift will be 118°. That is, the output will be 118° from the noninverting input, and 62° from the inverting. Thus, there is a 62° *phase margin* (180° − 118°) between input and output (assuming that the input signal is applied to the inverting input in the usual manner). A 62° phase margin should provide very stable operation.

Now assume that it is desired to operate at a higher frequency, but still provide the 60 dB gain. Follow the 60 dB line to intersect with the 300 pF curve (300 pF is the smallest recommended capacitor). The intersection occurs at approximately 600 kHz. However, the 600 kHz line intersects that phase curve at about 175°, resulting in a phase margin of about 5°. This will probably produce unstable operation. Thus, if 60 dB gain must be obtained, the capacitor value must be larger than 300 pF.

The curves of Fig. 1-22 show that, for a given gain, a larger value of phase-lead capacitance will reduce frequency capability, and vice versa. However, a reduction in frequency will increase stability. For a given frequency of operation, capacitor size has no effect on stability, only on gain.

Figure 1-23 shows the schematic of an IC *op-amp with built-in phase-lead capacitance.* The 30 pF capacitor C_1 between the collector of Q_{10} and the base of Q_{15} provides the desired phase shift. This results in a fixed rolloff at 6 dB/octave as shown in Fig. 1-24. Note that this rolloff approaches that of the "ideal" op-amp shown in Fig. 1-18. Any closed-loop gain (less than open-loop gain) can be selected by feedback resistances in the normal manner. Likewise, any combination of gain/frequency response can be selected. For example, if it is desired to start the rolloff at 10^4 Hz, choose a feedback resistance Z_F that is 10^2 times larger than the input resistance Z_R (such as $Z_F = 10,000$ and $Z_R \doteq 100$). The closed-loop gain is then flat to 10^4 Hz, and rolls off at 6 dB/octave to unity gain at 10^6 Hz.

Keep in mind that the gain/frequency characteristics of the op-amp in Fig. 1-23 cannot be changed. Thus, if a different gain/frequency response is required (say 10^2 gain at 10^5 Hz), the op-amp is unsuitable.

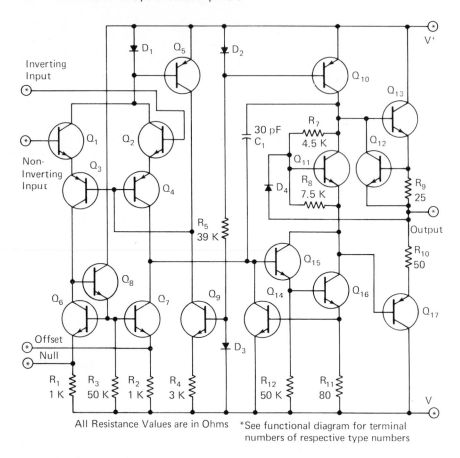

Fig. 1-23. Op-amp with internal phase compensation (Courtesy RCA).

Figure 1-25 shows a somewhat different system of phase-lead graphs as they appear on the manufacturer's data. These graphs show both the uncompensated and compensated characteristics on the same graph. The curves marked A are the uncompensated open-loop voltage gain (A_{VOL}) and phase shift. The B curves show compensated gain and phase shift.

When uncompensated, the gain starts the 6 dB/octave rolloff at about 0.3 MHz, the 12 dB/octave rolloff at about 5 MHz, and the 18 dB/octave rolloff at about 20 MHz. The phase shift exceeds 180° at about 4.5 MHz. Thus, the absolute maximum uncompensated operating frequency is about 4 MHz.

With compensation, 6 dB/octave rolloff starts at about 0.002 MHz. Unity gain occurs at about 4.5 MHz. The compensated phase shift never exceeds about 150°, which occurs at about 10 MHz. Thus, stable opera-

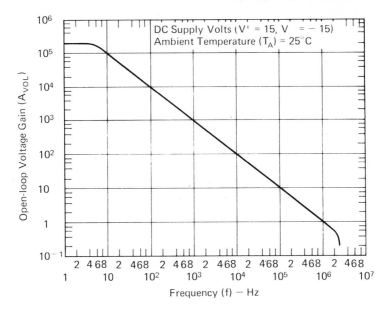

Fig. 1-24. Open-loop gain versus frequency characteristics of op-amp with internal phase compensation (Courtesy RCA).

tion is possible up to 10 MHz, even though there is no gain beyond about 4.5 MHz. As with the other examples, any combination of gain and frequency can be selected along the compensated A_{VOL} line by proper selection of feedback and input resistances.

1-2.8 Conventional phase-lag compensation examples

As shown in Fig. 1-26, conventional phase-lag compensation requires the addition of a capacitor and resistor to the basic op-amp circuit. This RC network is connected across a circuit component (such as across a stage output or input, or across the op-amp output). Conventional phase-lag compensation is external to the op-amp circuit. Many IC op-amps have terminals provided for connection of external phase compensation circuits to internal circuit.

As in the case of phase-lead compensation, the manufacturer's information must be used for phase-lag compensation. The only alternative is to use the equations shown in Fig. 1-26, and test the results. This is not recommended unless the manufacturer's data are not available. Always use the published information, at least as a first trial value.

As shown in Fig. 1-26, the values of the external phase compensating resistor and capacitor are dependent upon the frequency at which the

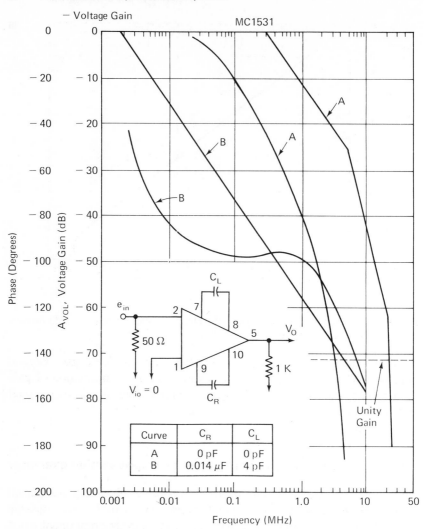

Fig. 1-25. Op-amp characteristics with and without phase-lead compensation (Courtesy Motorola).

uncompensated gain changes from 6 dB/octave to 12 dB/octave (generally known as the second pole of the uncompensated gain curve). Thus, it is relatively simple to find values for R and C if the uncompensated gain characteristics are known (or can be found by test). However, note that the frequency at which the compensated rolloff will start (the start of the compensated 6 dB/octave rolloff) is dependent upon the value of the external capacitor C and internal characteristics of the op-amp circuit. Thus, even though satisfactory values of R and C can be found to

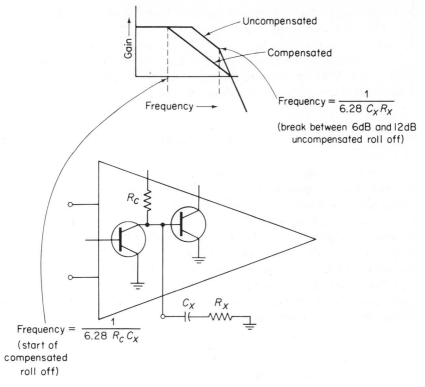

Fig. 1-26. Frequency response compensation with external capacitor and resistor (conventional phase lag).

produce a straight rolloff, there is no way of determining the frequency at which rolloff will start (using the equations).

If conventional phase-lag rolloff is to be used, and the values must be found by test (no manufacturer's data), assume a convenient value for R, and find C using the equations of Fig. 1-26. Then test the compensated op-amp. If the compensated rolloff starts at too low a frequency, decrease the value of C, and find a new corresponding value for R.

As an example, assume that the uncompensated rolloff changes from 6 dB/octave to 12 dB/octave at 10 MHz, and it is desired to have the compensated rolloff start at 300 kHz. Assume that R_x is 1000 ohms (a convenient value). Using the equation of Fig. 1-26, a frequency of 10 MHz, and an R_x of 1000 ohms, the value for C_x is:

$$C_x = \frac{1}{6.28 \times 10 \text{ MHz} \times 1000} \approx 16 \text{ pF}$$

Test the op-amp with C_x at 16 pF and R_x at 1000 ohms. If the compensated rolloff starts at some frequency lower than 300 kHz, increase the value of C_x and find a new value for R_x. If the rolloff starts above 300 kHz, decrease the value of C, and use another value for R.

For example, if compensated rolloff starts at 100 kHz instead of the desired 300 kHz, increase the value of C_x to 30 pF. Using the equation of Fig. 1-26, the corresponding value of R_x is:

$$R_x = \frac{1}{6.28 \times 10 \text{ MHz} \times 30 \text{ pF}} \approx 530 \text{ ohms}$$

Figures 1-27 and 1-28 show typical phase-lag compensation characteristics for an op-amp. Figure 1-27 shows the uncompensated gain characteristics (curve A), the phase-lag compensated characteristics (curve B), and phase-lead characteristics (curve C). Note that the uncompensated gain changes from a 6 dB/octave rolloff to a 12 dB/octave rolloff at about 10 MHz. Thus, 10 MHz is used in the equation to find the values of R and C. As shown, the recommended values are 18 pF and 820 ohms. Rolloff starts at some frequency below 0.1 MHz, and continues at 6 dB/octave down to unity gain at about 33 MHz, when the recommended RC combination is used.

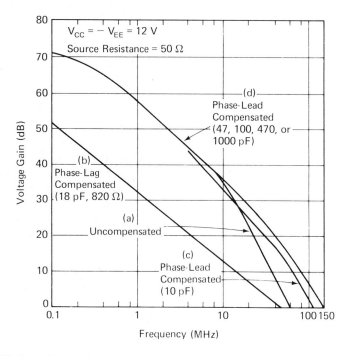

Fig. 1-27. Open-loop gain as a function of frequency for compensated and uncompensated op-amps (Courtesy RCA).

Although the phase-lag capacitance of 18 pF shown in curve B of Fig. 1-27 is sufficient to provide stable operation (no oscillation) in a resistance feedback system down to unity gain, the capacitance is not sufficient to provide a flat, closed-loop response (flat to within ± 1 dB) below 20 dB. That is, if the feedback network is chosen to provide something less than 20 dB closed-loop gain, operation will be stable, but the response will not be flat (there will be peaking or shifts in gain greater than ± 1 dB).

This condition is shown in Fig. 1-28. For example, if the closed-loop gain is set for 10 dB, the recommended capacitance value is 56 pF. For unity gain, the capacitance value must be increased to 68 pF. A corresponding value of R must be found in each case. For example, if 56 pF is used for 10 dB gain, R must be reduced to about 280 ohms. If C is made 68 pF for unity gain, R is reduced to about 230 ohms.

1-2.9 Miller-effect phase-lag compensation examples

As shown in Fig. 1-29, the problems in determining Miller-effect phase-lag values are essentially the same as for conventional phase-lag. That is, the values of the compensating resistor and capacitor are dependent upon the frequency at which uncompensated rolloff changes from 6 to 12 dB. The frequency at which compensated rolloff starts is dependent upon compensating RC values, and internal op-amp values. (This is also true for the frequency at which compensated rolloff reaches unity gain.)

Again, always use manufacturer's information. If Miller-effect phase compensation values must be found without manufacturer's data, assume a convenient value for R, and find C using the equations of Fig. 1-29. Decrease the value of C if compensated rolloff starts at too low a frequency, and vice versa.

Figure 1-30 illustrates typical Miller-effect phase-lag compensation characteristics for an op-amp. Note that the values of phase-compensating resistors and capacitors were found by manufacturer's test, and differ considerably from the theoretical values of Fig. 1-29.

1-3. OPERATIONAL AMPLIFIER CHARACTERISTICS

The user must have a thorough knowledge of op-amp characteristics to get the best possible results from the op-amp in any system. From a user's standpoint, op-amp characteristics provide a good basis for op-amp system design. However, commercial op-amps, either IC

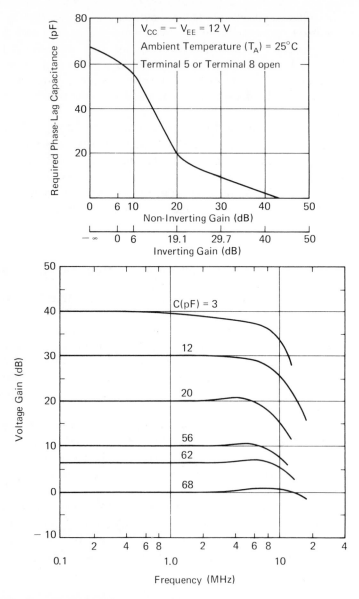

Fig. 1-28. Amount of phase-lag capacitance required to obtain a flat (± 1 dB) response, and typical response characteristics (Courtesy RCA).

or discrete component, are generally designed for specific applications. Although some commercial op-amps are described as "general purpose," it is impossible to design an op-amp with truly universal characteristics.

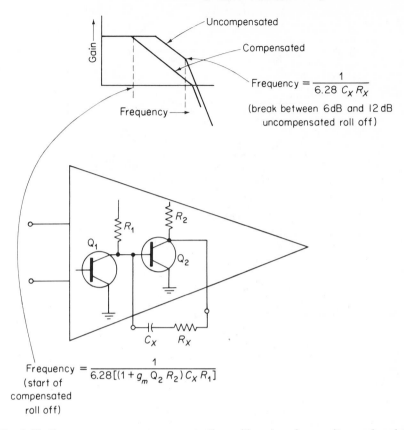

Fig. 1-29. Frequency response compensation with external capacitor and resistor (Miller-effect rolloff).

For example, certain op-amps are designed to provide high-frequency gain at the expense of other performance characteristics. Other op-amps provide very high gain or high input impedance in low-frequency applications. IC op-amps, which are fabricated by the diffusion process, can be made suitable for comparator applications (where both halves of the differential amplifier circuits must be identical). Likewise, IC op-amps can be processed to provide high gain at low dissipation levels. For these reasons, any description of op-amp characteristics must be of a general nature, unless a specific application is being considered.

Most of the op-amp characteristics required for proper use of the op-amp in any application can be obtained from the manufacturer's datasheet or similar catalog information. There are some exceptions to this rule. For certain applications it may be necessary to test the op-amp under simulated operating conditions. Test procedures for measurement of characteristics discussed here are described in Chapter 5.

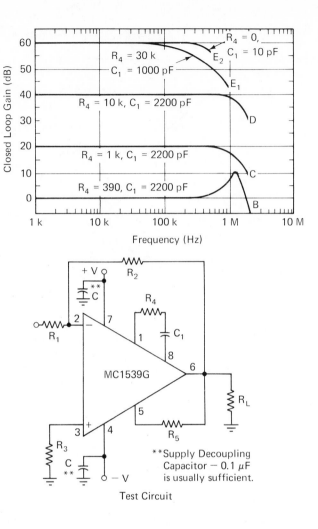

Recommended Test Conditions

Curve No.	Voltage Gain	Test Conditions				
		R_1 (Ω)	R_2 (Ω)	R_3 (Ω)	R_4 (Ω)	C_1 (pF)
A	A_{VOL}	0	∞	0	∞	0
B	1	10 k	10 k	5 k	390	2200
C	10	1 k	10 k	1 k	1 k	2200
D	100	1 k	100 k	1 k	10 k	2200
E_1	1000	1 k	1 M	1 k	30 k	1000
E_2	1000	1 k	1 M	1 k	0	10

Fig. 1-30. Typical Miller-effect phase-lag compensation characteristics (Courtesy Motorola).

46

In using datasheet information or test results, or both, it is always necessary to interpret the information. Each manufacturer has its own system of datasheets. It is impractical to discuss all datasheet formats here. Instead, we shall discuss typical information found on op-amp datasheets, as well as test results, and see how this information affects the op-amp user.

1-3.1 Open-loop voltage gain

The open-loop voltage gain (A_{VOL} or A_{OL}) is defined as the ratio of a change in output voltage to a change in input voltage at the input of the op-amp. Open-loop gain is always measured without feedback and usually without phase-shift compensation.

Open-loop gain is *frequency dependent* (gain decreases with increased frequency). This is shown in Fig. 1-31. As shown, the gain is flat (within about ± 3 dB) up to frequencies of about 0.1 MHz. Then the gain rolls off to unity at frequencies above 10 MHz. The open-loop gain is also *temperature dependent,* and dependent upon supply voltage, as shown in Fig. 1-31. Generally, gain increases with supply voltage. The effects of temperature on gain are different at different frequencies.

Ideally, open-loop gain should be infinitely high since the primary function of an op-amp is to amplify. In general, the higher the gain the higher the accuracy of op-amp transfer function (relationship of output

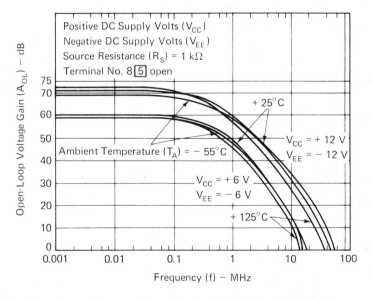

Fig. 1-31. Open-loop gain versus frequency for CA3008 (Courtesy RCA).

to input). However, there are practical limits to gain magnitude and also levels at which an increase in magnitude buys little in the way of increased performance. The true significance of open-loop gain is many times mis-applied in op-amp operation where in reality open-loop gain determines closed-loop accuracy limits, rather than ultimate accuracy.

The numerical values of the open-loop gain (and the bandwidth) of an op-amp are of relatively little importance in themselves. The important requirement is that the open-loop gain must be greater than the closed-loop gain over the frequency of interest if an accurate transfer function is to be maintained. For example, if a 40 dB op-amp and a 60 dB op-amp are used in a 20 dB closed-loop gain configuration, and the open-loop gain is decreased 50 percent in each case (say due to component aging), the closed-loop gain of the 40 dB op-amp varies 9 percent, and that of the 60 dB op-amp varies only 1 percent.

The *frequency rolloff characteristics* are the prime determinants of op-amp frequency response. The greater the rate of rolloff prior to the intersection of the feedback ratio (closed-loop) frequency characteristics with the open-loop response (in the active region), the more difficult phase compensation of the op-amp becomes.

An 18 dB/octave rolloff is generally considered the maximum slope that can occur in the active region before proper phase compensation becomes extremely difficult or impossible to achieve. In addition, be-cause op-amps have useful application down to and including unity gain, the active region of the op-amp may be considered as the entire portion of the frequency characteristic above the 0 dB bandwidth. Thus, a well-designed op-amp should roll off at no greater than 18 dB/octave until well below unity gain.

As discussed, open-loop gain can be modified by several compensation methods. A typical op-amp datasheet will show the results of such com-pensation, usually by means of graphs such as the one shown in Fig. 1-25.

After compensation is applied, the op-amp can be connected in the closed-loop configuration. The voltage gain under closed-loop conditions is dependent upon external components (the ratio of feedback resistance to input resistance). Thus, closed-loop gain is usually not listed as such on op-amp datasheets. However, the datasheet may show some typical gain curves with various ratios of feedback (Fig. 1-30 is an example). If available, such curves can be used directly to select values of feedback components (as well as phase compensation components).

1-3.2 Phase shift

Figure 1-32 shows the open-loop phase shift curve of a typical op-amp. (Sometimes, such curves are included on the open-loop frequency response graph.) Because a closed-loop gain of unity allows the highest frequency response for the loop gain, the closed-loop unity gain frequency is considered the worst case for phase shift.

One figure of merit commonly used in evaluating the stability of an op-amp is *phase margin*. As discussed in Sec. 1-2, oscillations can be sustained if the total phase shift around the loop (from input to output, and back to input) can reach 360° before the total gain around the loop drops below unity (as the frequency is increased). Because an op-amp is normally used in the inverting mode, 180° of phase shift is available to begin with. Additional phase shift is developed by the op-amp due to internal circuit conditions.

Phase margin represents the difference between 180° and the phase shift introduced by the op-amp at the frequency at which loop-gain is unity. A value of 45° phase margin is considered quite conservative to provide a guard against production variations, temperature effects and other stray factors. This means that the op-amp should not be operated at a frequency at which the phase shift exceeds 135° (180° − 45°). However, it is possible to operate op-amps at frequencies at which the phase shift is kept within the 160° to 170° region. Of course, the ultimate stability of an op-amp must be established by tests.

1-3.3 Bandwidth, slew rate and output characteristics

The bandwidth, slew rate, output voltage swing, output current and output power of an op-amp are all interrelated. These characteristics are frequency dependent, and depend upon phase compensation. The characteristics are also temperature and power supply depen-

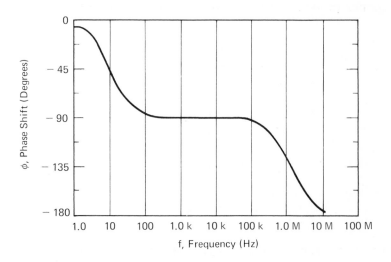

Fig. 1-32. Open-loop phase shift for MC1556 (Courtesy Motorola).

dent, but to a lesser extent. Before discussing the interrelationship, let us define each of the characteristics.

Bandwidth for an op-amp is usually expressed in terms of open-loop operation. The common term is BW_{OL} at −3dB, such as shown in Fig. 1-33. For example, a BW_{OL} of 800 kHz indicates that the open-loop gain of the op-amp will drop to a value of 3 dB below the flat or low-frequency level at a frequency of 800 kHz.

Frequency range is sometimes used in place of open-loop bandwidth. The frequency range of an op-amp is often listed as "useful frequency range" (such as dc up to 18 MHz). Useful frequency range for an op-amp is similar to the F_T (total frequency) term used with discrete transistors. Generally, the high-frequency limit specified for an op-amp is the frequency at which gain drops to unity.

Power bandwidth is a more useful characteristic since it represents the bandwidth of the op-amp in closed-loop operation connected to a normal load. As shown in Fig. 1-34, power bandwidth is given as the peak-to-peak output voltage capability of the op-amp (working into a given load) across a band of frequencies. Power bandwidth figures usually imply that the output indicated is free of distortion, or that distortion is within limits (such as total harmonic distortion less than 5 percent). In the op-amp of Fig. 1-34, the output voltage is about 27V (p-p) up to about 20 kHz, and drops to near zero at 300 kHz.

Power output of an op-amp is generally listed in terms of power across a given load (such as 250 mW across 500 ohms). However, power output is usually listed at only one frequency. The same is true of *output current,* or *maximum output current* characteristics found on some datasheets. Thus, power bandwidth is the more useful characteristic.

Output voltage swing is defined as the peak or peak-to-peak output

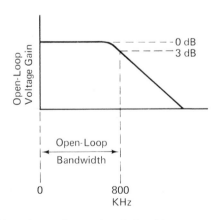

Fig. 1-33. Bandwidth and open-loop gain relationships.

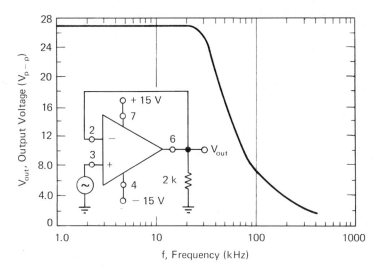

Fig. 1-34. Power bandwidth for MC1556 (Courtesy Motorola).

voltage swing (referred to zero) that can be obtained without clipping. A symmetrical voltage swing is dependent upon frequency, load current, output impedance and slew rate. Generally, an increase in frequency will decrease the possible output voltage swing. Figure 1-35 shows that maximum output voltage drops as frequency is increased. Also note that phase compensation capacitance has an effect on maximum output voltage. Figure 1-36 shows that an increase in load resistance will also increase maximum output voltage.

The *slew rate* of an op-amp is the maximum rate of change of the output voltage, with respect to time, that the op-amp is capable of producing while maintaining linear characteristics (symmetrical output without clipping).

Slew rate is expressed in terms of:

$$\frac{\text{difference in output voltage}}{\text{difference in time}} \text{ or } \frac{dV_o}{dt}$$

Usually, slew rate is listed in terms of *volts per microsecond*. For example, if the output voltage from an op-amp is capable of changing 7V in 1 μS, then the slew rate is 7. If, after compensation or other change, the op-amp changes a maximum of 3V in 1 μS, the new slew rate is 3.

Slew rate of an op-amp is the direct function of the phase-shift compensation capacity. At higher frequencies, the current required to charge and discharge a compensating capacitor can limit available current to succeeding stages or loads, and thus result in lower slew rates. This is

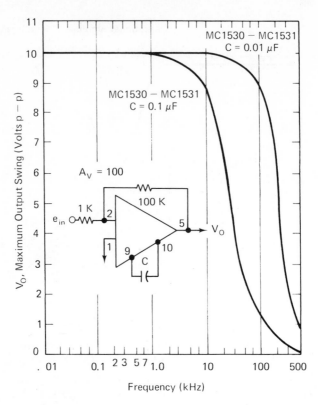

Fig. 1-35. Maximum output voltage swing versus frequency (Courtesy Motorola).

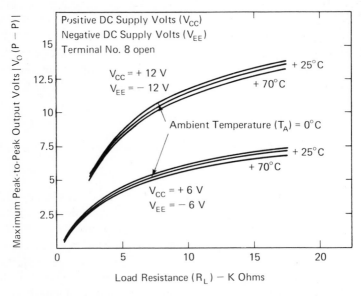

Fig. 1-36. Maximum peak-to-peak output voltage versus load resistance (Courtesy RCA).

one reason why op-amp datasheets usually recommend the compensation of early stages in the op-amp where signal levels are still small and little current is required.

Slew rate decreases as compensation capacitance increases. This is shown in Fig. 1-37. Thus, where high frequencies are involved, the lowest values of compensation capacitor should be used. Figure 1-38 shows the minimum capacitance values that can be used with different closed-loop gain levels for the particular op-amp. The curves of Figs. 1-37 and 1-38 are typical of those found on op-amp datasheets in which slew rate is of particular importance.

The major effect of slew rate in op-amp applications is on output power. All other factors being equal, a lower slew rate results in lower power output. Slew rate and the term *full power response* of an op-amp are directly related. Full power response is the maximum frequency measured in a closed-loop unity gain configuration for which rated output voltage can be obtained for a sinewave signal, with a specified load, and without distortion due to slew rate limiting.

The slew rate versus full power response relationship can be shown as:

$$\text{slew rate (in volts/second)} = 6.28 \times F_M \times E_0$$

where F_M is the full power response frequency (in Hz), and E_0 is the peak output voltage (one-half the peak-to-peak voltage).

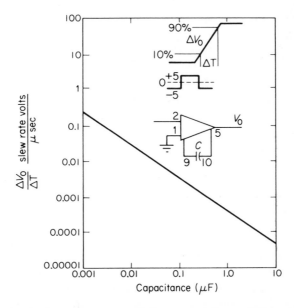

Fig. 1-37. Slew rate of op-amp versus rolloff compensation capacitance.

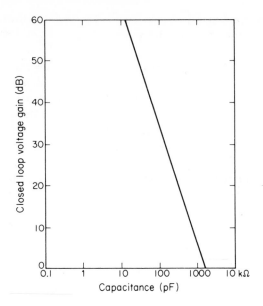

Fig. 1-38. Closed-loop voltage gain of op-amp versus minimum rolloff capacitance.

For example, using the characteristics shown in Fig. 1-34, the output voltage E_o is about $13\,V$ (one-half the peak-to-peak of $26\,V$) at a frequency of 30 kHz. Thus, the slew rate is:

$$\text{slew rate} = 6.28 \times 30{,}000 \times 13 \approx 2{,}449{,}200\,V/S \approx 2.45\,V/\mu S$$

The equation can be turned around to find the full power response frequency. For example, assume that an op-amp is rated as having a slew rate of $2.5\,V/\mu S$ and a peak-to-peak output of $20\,V$ ($E_o = 10\,V$). Find the full power response frequency F_M as follows:

$$F_M = \frac{2.5\,V/\mu S}{6.28 \times 10} = \frac{2{,}500{,}00\ V/S}{62.8} \approx 40{,}000\ Hz \approx 40\ kHz$$

Of course, if curves such as shown in Fig. 1-34 are available, it is not necessary to calculate the maximum frequency for a given output. Simply follow the 20V line until it crosses the curve at 40 kHz.

The graph of Fig. 1-39 shows the relationship between slew rate, full power response frequency and output voltage. For example, if slew rate is $6\,V/\mu S$, the maximum output voltage (peak-to-peak) is about 20V at a frequency of 100 kHz, and vice versa. If slew rate is $10\,V/\mu S$, the peak-to-peak output is increased to 30V at 100 kHz, and vice versa.

With a constant output load, the power output of an op-amp is de-

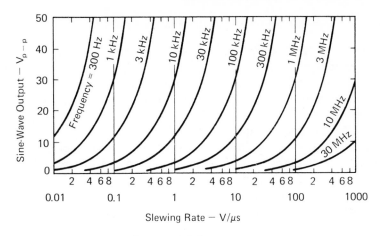

Fig. 1-39. Slewing rate curve (Courtesy RCA).

pendent upon output voltage. In turn, all other factors being equal, output voltage is dependent upon the slew rate. Since slew rate depends upon phase compensation capacitance, op-amp power output is also dependent upon compensation. Some datasheets omit slew rate, but provide a graph similar to that shown in Fig. 1-40. This graph shows the direct relationship between full power output frequency and phase compensation capacitance. For example, with a phase compensation capacitance of 0.01 μF and a 500 ohm load, the op-amp shown in Fig. 1-40 will deliver full-rated output power up to a frequency of 80 kHz.

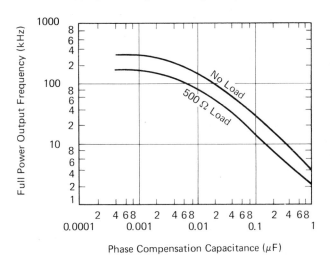

Fig. 1-40. Frequency for full power output as a function of phase-compensating capacitance.

1-3.4 Input and output impedance

Input impedance is defined as the impedance seen by a source looking into one input of the op-amp with the other input grounded (see Fig. 1-41). The primary effect of input impedance on design is to reduce amplifier loop gain. If the input impedance is quite different from the impedance of the device driving the op-amp (source impedance), there will be a loss of input signal due to the mismatch. However, in practical terms, it is not possible to alter the op-amp impedance. Thus, if impedance match is critical, either the op-amp or driving source must be changed to effect a match.

Output impedance is defined as the impedance seen by a load at the output of the op-amp (see Fig. 1-42). Excessive output impedance can reduce the gain since, in conjunction with the load and feedback resistors, output impedance forms an attenuator network. In general, output impedance of op-amps is less than 200 ohms. Generally input resistances are at least 1000 ohms, with feedback resistance several times higher than 1000 ohms. Thus, the output impedance of a typical op-amp will have little effect on gain.

If the op-amp is serving primarily as a voltage amplifier (as is usually

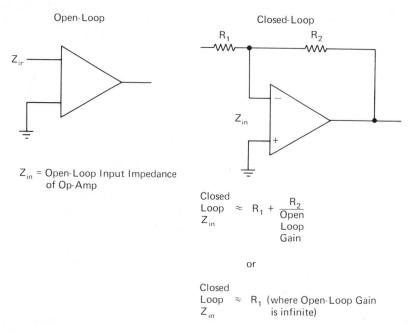

Open-Loop

Closed-Loop

Z_{ir}

Z_{in} = Open-Loop Input Impedance of Op-Amp

R_1 R_2

Z_{in}

$$\text{Closed Loop } Z_{in} \approx R_1 + \frac{R_2}{\text{Open Loop Gain}}$$

or

$$\text{Closed Loop } Z_{in} \approx R_1 \text{ (where Open-Loop Gain is infinite)}$$

Fig. 1-41. Input impedance relationships.

the case) the effect of output impedance will be at a minimum. Output impedance has a more significant effect in design of power applications where the op-amp must supply large amounts of load current.

Closed-loop output impedance is found by using the equation of Fig. 1-42. Thus, it will be seen that output impedance will increase as frequency increases, since open-loop gain decreases.

Both input and output impedance will change with temperature, as well as frequency. This is shown in Fig. 1-43. Generally, both characteristics are listed on the datasheet at 25°C and 1 kHz, unless a graph is provided.

The ideal values for the input and output impedances of an op-amp are infinity and zero, respectively. The degree to which a practical op-amp approximates these values depends, for the most part, upon the application. A 5000 ohm open-loop input impedance may be quite sufficient for one application, whereas a 0.1 megohm input impedance may not be sufficient for another.

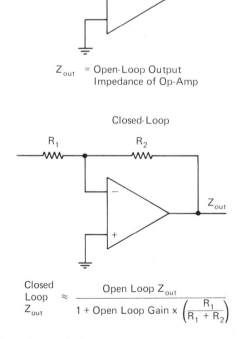

Open-Loop

Z_{out} = Open-Loop Output
Impedance of Op-Amp

Closed-Loop

$$\text{Closed Loop } Z_{out} \approx \frac{\text{Open Loop } Z_{out}}{1 + \text{Open Loop Gain} \times \left(\dfrac{R_1}{R_1 + R_2}\right)}$$

Fig. 1-42. Output impedance relationships.

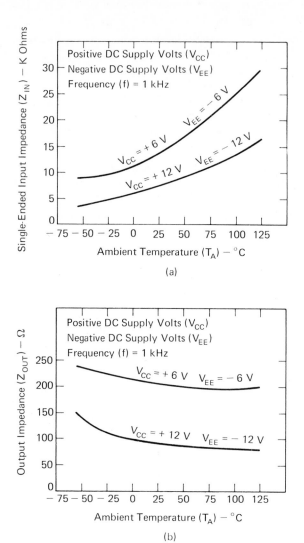

Fig. 1-43. Input and output impedance versus temperature for CA3008 (Courtesy RCA).

1-3.5 Input common-mode voltage swing

Input common-mode voltage (V_{ICM}) is defined as the maximum peak input voltage that can be applied to either input terminal of the op-amp without causing abnormal operation or damage (see Fig. 1-44). Some op-amp datasheets list a similar term: *common mode input voltage range* (V_{CMR}). Usually, V_{ICM} is listed in terms of peak voltage, with positive or negative peaks being equal. V_{CMR} is often listed for positive and negative voltages of different value (such as $+1V$ and $-3V$).

In practical use, either of these parameters limits the differential signal amplitude that can be applied to the op-amp input. So long as the input signal does not exceed the V_{ICM} or V_{CMR} values (in either the positive or negative directions), there should be no problem.

Note that some op-amp datasheets list "single-ended" input voltage signal limits where the differential input is not to be used.

1-3.6 Common-mode rejection ratio

Common-mode rejection terms are defined in Sec. 1-1.2. Figure 1-45 shows common-mode characteristics for a typical op-amp. Figure 1-45a shows that there is almost 80 dB difference between common-mode gain and differential gain. Figure 1-45b shows that common mode rejection ratio is almost 30 dB greater than differential gain.

Under differential drive conditions, the common-mode rejection has no drastic effects on the performance of an op-amp, unless the rejection ratio is extremely low. However, in a common-mode drive application, such as in a comparator, high common-mode rejection is essential. For example, if an op-amp with a 60 dB differential gain, and a 50 dB common-mode rejection, is used to compare a 1 volt signal against a 1 volt reference, the output will be 3.2 volts, when it should be zero. Such results would be totally unacceptable for a comparator type application. This is why the common-mode rejection should be at least 20 dB greater than the differential gain.

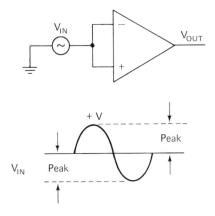

Fig. 1-44. Input common-mode voltage swing relationships.

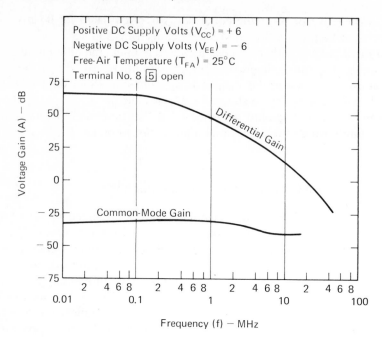

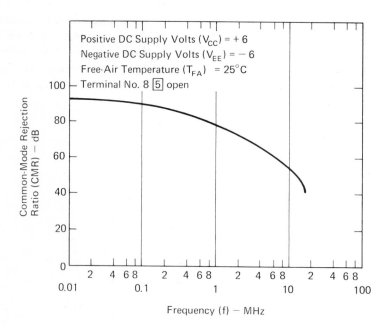

Fig. 1-45. Common-mode rejection characteristics (Courtesy RCA).

1-3.7 Input bias current

Input bias current is defined as the average value of the two input bias currents of the op-amp differential input stage. This is shown in Fig. 1-46, which illustrates input bias for two typical IC op-amps. Note that input bias current decreases as temperature increases. Input bias current is essentially a function of the large signal current gain of the input stage.

In use, the significance of input bias current is the resultant voltage drop across input resistors or other source resistances. This voltage drop can restrict the input common-mode voltage range at higher impedance levels. The voltage drop must be overcome by the input signal. Also, a large input bias current is undesirable in applications where the source cannot accommodate a significant dc current. Examples of such applications are those in which the source resistance is very large (resulting in a large voltage drop), or sources of a magnetic nature that can be

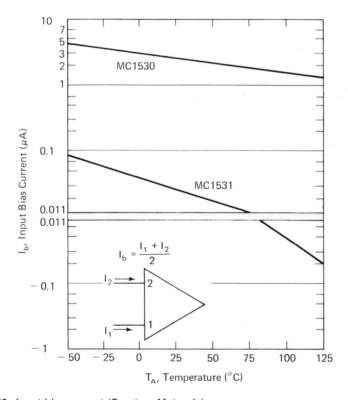

Fig. 1-46. Input bias current (Courtesy Motorola).

severely unbalanced by a flow of dc current (such as transducers that operate on magnetic principles).

Some op-amps have very low input bias current, and are thus well suited to these applications. Where very low input bias current is required, the input differential stages of the op-amp often use field-effect transistors or other MOS devices that draw very little input current.

1-3.8 Input offset voltage and current

Input offset voltage is defined as the voltage that must be applied at the input terminals to obtain zero output voltage (see Fig. 1-47). Input offset voltage indicates the matching tolerance in the differential amplifier stages. A perfectly matched amplifier requires zero input voltage to provide zero output voltage. Typically, input offset voltage is on the order of 1 or 2 mV for an IC op-amp, as shown in Fig. 1-47a. The offset voltage for an op-amp can also be defined as the deviation of the output dc level from the arbitrary input-output level, usually taken as ground reference when both inputs are shorted together.

Input offset current is defined as the difference in input bias current into the input terminals of an op-amp (Fig. 1-47c). Input offset current is an indication of the degree of matching of the input differential stage. Typically, input offset current is on the order of 1 or 2 μA for an IC op-amp, as shown in Fig. 1-47a. The offset current for an op-amp can also be defined as the deviation when the inputs are driven by two identical dc input bias current sources.

Offset voltage and current are usually referred back to the input because their output values are dependent upon feedback. (That is, datasheets rarely list output offset characteristics.) In normal use, the offset in an op-amp results from a combination of offset voltage and current. For example, if an op-amp has a 1 mA input offset voltage, and a 1 μA input offset current, with the inputs returned to ground through 1000 ohm resistors, the total input offset is either zero or 2 mV, depending upon the phase relationship between the two offset characteristics.

The offset of an op-amp is a direct-current error that should be minimized for numerous reasons, including the following:

1. The use of an op-amp as a true dc amplifier is limited to signal levels much greater than the offset;
2. comparator applications require that the output voltage be zero (within limits) when the two input signals are equal and in phase;
3. in a dc cascade the offset of the first stage determines the offset characteristics of the entire system.

Thus, any offset at the input of an op-amp is multiplied by the gain at the output. If the op-amp serves to drive additional amplifiers, the increased offset at the op-amp output will be multiplied even further. The gain of the entire system must then be limited to a value that is insufficient to cause limiting in the final output stage.

The effect of input offset voltage on op-amp use is that the input signal must overcome the offset voltage before an output will be produced. For example, if an op-amp has a 1 mV input offset voltage, and a 1 mV signal is applied, there is no output. If the signal is increased to 2 mV, the op-amp will produce only the peaks. Since input offset voltage is increased

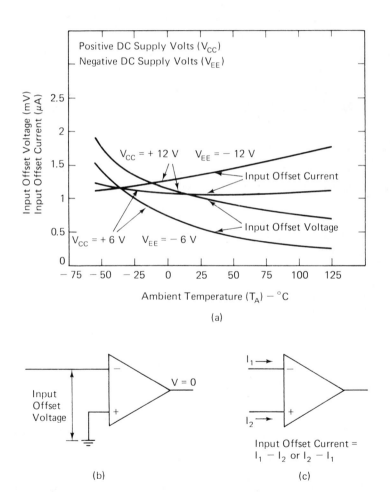

(a)

(b)   (c)

Fig. 1-47. Input offset voltage and current.

by gain, the effect of input offset voltage is increased by the ratio of feedback resistance to input resistance plus unity (or one) in the closed loop condition. For example, if the ratio is 100 to 1 (for a gain of 100), the effect of input offset voltage is increased by 101.

Input offset current can be of greater importance than input offset voltage when high impedances are used in design. If the input bias current is different for each input, the voltage drops across the input resistors (or input impedance) will not be equal. If the resistance is large, there will be a large unbalance in input voltages. This condition can be minimized by means of a resistance connected between the noninverting input and ground, as shown in Fig. 1-48. The value of this resistor R_3 should equal the parallel equivalent of the input and feedback resistors R_1 and R_2, as shown by the equations. In practical design, the trial value for R_3 is based on the equation of Fig. 1-48. The value of R_3 is then adjusted for minimum voltage difference at both terminals (under normal operating conditions, but with no signal).

Some op-amps (particularly IC op-amps) include provisions to neutralize any offset. Typically, an external voltage is applied through a potentiometer to terminals on the op-amp. The voltage is adjusted until the offset, at the input and output, is zero. For example, note the terminals marked "offset null" on the schematic of Fig. 1-23. The terminals are connected to the emitters of the first differential amplifier stage. Figure 1-49 shows the external offset null or neutralization circuit used with the op-amp of Fig. 1-23.

For op-amps without offset compensation, the effects of input offset can be minimized by an external circuit. Figure 1-50 shows two such circuits, one for inverting and the other for noninverting op-amps. The

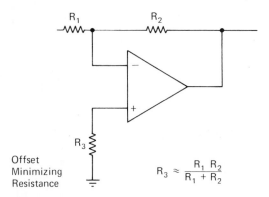

Fig. 1-48. Minimizing input offset current (and input offset voltage) by means of resistor at non-inverting input.

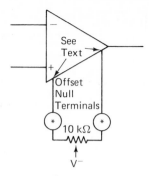

Fig. 1-49. Typical offset null or neutralization circuit.

equations shown in Fig. 1-50 assume that resistor R_B must be of a value to produce a null range of ± 7.5 mV. This is generally sufficient for any op-amp. However, if a different input offset voltage range is required, simply substitute the desired range for ±7.5 mV.

In addition to the basic circuits of Fig. 1-50, some of the applications described in Chapters 2 through 4 require special offset null circuits. One reason for an offset null is that the input and output dc levels of an op-amp should be equal, or nearly equal. This condition is desirable to assure that the resistive feedback network can be connected between the input and output without upsetting either the differential or the common-mode dc bias.

The average temperature coefficient of input offset voltage, listed on some datasheets as TCV_{IO}, is dependent upon the temperature coefficients of various components within the op-amp. Temperature changes affect stage gain, match of differential amplifiers and so forth, and thus change input offset voltage. From a user's standpoint, TCV_{IO} need be considered only if the parameter is large, and the op-amp must be operated under extreme temperatures. For example, if input offset voltage doubles with an increase to a temperature that is likely to be found during normal operation, the higher input offset voltage should be considered the "normal" value for design.

1-3.9 Power supply sensitivity
(input offset voltage sensitivity)

Power supply sensitivity is defined as the ratio of change in input offset voltage to the change in supply voltage producing it, with the remaining supply held constant (See Fig. 1-51). Some op-amp data-sheets list a similar characteristic: *input offset voltage sensitivity*. In

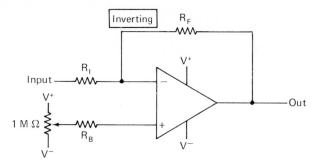

Value of R_B required to have a null adjustment range of ± 7.5 mV:

$$R_B \approx \frac{R_I V^+}{7.5 \times 10^{-3}} \quad \text{assuming } R_B \gg R_I$$

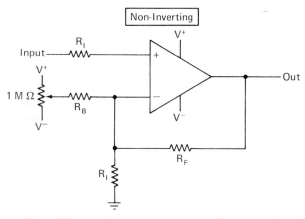

Value of R_B required to have a null adjustment range of ± 7.5 mV:

$$R_B \approx \frac{R_I R_F V^+}{(R_I + R_F)\, 7.5\,\text{mV} \times 10^{-3}}$$

$$\text{Assuming } R_B \gg \frac{R_I R_F}{R_I + R_F}$$

Fig. 1-50. Input offset minimizing circuits.

either case, the characteristic is expressed in terms of mV/V or μV/V, representing the change (in mV or μV) of input offset voltage to a change (in volts) of one power supply. Usually there is a separate sensitivity characteristic for each power supply, with the power supply assumed to be held constant. For example, a typical listing is 0.1 mV/V for a positive supply. This implies that with the negative supply held constant, the input offset voltage will change by 0.1 mV for each 1V change in positive supply voltage.

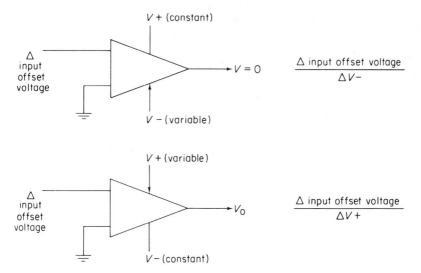

Fig. 1-51. Power supply sensitivity.

The effects of power supply sensitivity (or input offset voltage sensitivity) are obvious. If an op-amp has considerable sensitivity to power supply variations, overall performance is affected by each supply voltage change. The power supply regulation must be increased to provide correct operation with minimum input signal levels.

1-3.10 Noise voltage

There are many systems for measuring noise voltage in an op-amp, and equally as many methods used to list the value on datasheets. Some datasheets omit the value entirely. In general, noise is measured with the op-amp in the open-loop condition, with or without compensation, and with the input shorted or with a fixed resistance load at the input terminals.

The input and/or output voltage is measured with a sensitive voltmeter or oscilloscope. Input noise is on the order of a few microvolts; output noise is usually less than 100 mV. Output noise is almost always greater than input noise (because of the amplifier gain).

Except in cases where the noise value is very high or the input signal is very low, amplifier noise can be ignored (unless a particular application requires extremely low noise). Obviously, a 10 μV noise at the input will mask a 10 μV signal. If the signal is raised to 1 mV with the same op-amp, the noise will be unnoticed.

Noise is dependent upon temperature as well as upon the method of compensation used. This is shown in Fig. 1-52, which illustrates noise

characteristics for typical IC op-amps. Since op-amp noise characteristics are related directly to the method used to test the values, a more
thorough discussion of noise figure is given in Sec. 5-11 of Chapter 5.

1-3.11 Power dissipation

An IC op-amp datasheet usually lists two power dissipation ratings. The same is true of some discrete component op-amps. One

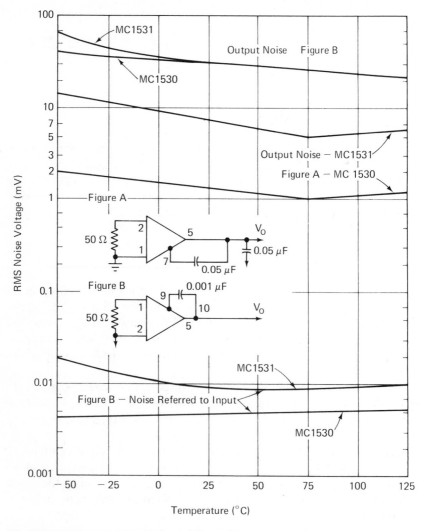

Fig. 1-52. RMS noise voltage (Courtesy Motorola).

value is the *total device dissipation*, which includes any load current. The other value is *device dissipation*, which is defined as the dc power dissipated by the op-amp itself (with output at zero and no load).

The device dissipation must be subtracted from the total dissipation to calculate the load dissipation.

For example, if an IC op-amp can dissipate a total of 300 mW (at a given temperature and supply voltage, and with or without a heat sink) and the IC op-amp itself dissipates 100 mW, the load cannot exceed 200 mW (300 − 100 = 200).

2. LINEAR APPLICATIONS

In this chapter, we shall discuss linear applications for op-amps. These linear applications include those cases in which input and output are essentially sinewaves, even though modified by the op-amp or the associated circuit. Nonlinear applications are discussed in Chapter 4. For the readers' convenience, the same format is used for each application (where practical).

First, a working schematic is presented for the circuit, together with a brief description of its function. Where practical, the working schematic also includes the operational characteristics of the circuit (in equation form), as well as rule-of-thumb relationships of circuit values (also in equation form).

Next, design considerations, such as desired performance, used with external circuits, amplification, operating frequency and so forth, are covered. This is followed by reference to equations (on the working schematic) and procedures for determining external component values that will produce the desired results.

Finally, a specific design problem is stated and a design example is given. The value of each external circuit component is calculated in step-by-step procedures, using guidelines established in the design considerations and/or working schematic equations.

The reader will note that the power supply and phase compensation connections are omitted from the schematics related to applications described here, except in Secs. 2-1 and 2-2. In all of the remaining applications it is assumed that the op-amp is connected to a power source as described in Sec. 2-1. Likewise, it is assumed that a suitable phase compensation scheme has been selected for the op-amp, as discussed in Sec. 2-2.

Unless otherwise stated, all of the design considerations for the basic op-amp described in Chapter 1 apply to each application covered here.

Where applicable, reference is made to Chapter 5 in which procedures for testing the completed circuit are given.

2-1. OP-AMP POWER SUPPLIES

Typically, op-amps require connection to both a positive and negative power supply. This is because most op-amps use one or more differential amplifiers. When two power supplies are required, the supplies are usually equal or symmetrical (such as +6V and −6V, +12V and −12V, etc.). This is the case with the op-amp of Fig. 2-1, which normally operates with +12V and −12V. It is possible to operate an op-amp that normally requires two supplies from a single supply by means of special circuits (external to the op-amp). Such circuits are discussed in Sec. 2-1.6.

2-1.1 IC op-amp power connections

Unlike most discrete transistor circuits in which it is usual to label one power supply lead positive and the other negative without specifying which (if either) is common to ground, it is necessary that all IC op-amp power supply voltages be referenced to a common or ground (which may or may not be physical or equipment ground).

As in the case of discrete transistors, manufacturers do not agree on power supply labeling for IC op-amps. For example, the circuit shown in Fig. 2-1 uses $V+$ to indicate the positive voltage and $V-$ to indicate

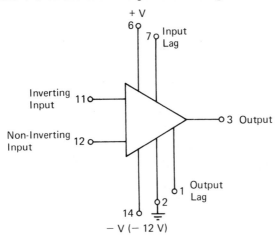

Fig. 2-1. Typical linear op-amp operating with symmetrical 12 V power supplies.

the negative voltage. Another manufacturer might use the symbols V_{EE} and V_{CC} to represent negative and positive, respectively. As a result, the op-amp datasheet must be studied carefully *before* applying any power source.

2-1.2 Typical op-amp power supply connections

Figure 2-2 shows typical power supply connections for an op-amp. The protective diodes shown are recommended for any power supply circuit in which the leads could be accidently reversed. The diodes permit current flow only in the appropriate direction. The op-amp of Fig. 2-2 requires two power sources (of 12V each) with the positive lead of one and the negative lead of the other tied to ground or common.

The two capacitors shown in Fig. 2-2 provide for decoupling (signal bypass) of the power supply. Usually, disc ceramic capacitors are used. The capacitors should always be connected as close to the op-amp terminals as is practical, not at the power supply terminals. This is to diminish the effects of lead inductance. It is particularly important to decouple each op-amp where two or more op-amps are sharing a common voltage supply. A guideline for op-amps power supply decoupling capacitors is to use values between 0.1 and 0.001 μF.

In addition to the capacitors shown in Fig. 2-2, some op-amp layouts may require additional capacitors on the power lines. The main problem with op-amp power supply connections is undesired oscillation due to feedback. Most modern op-amps are capable of producing high gain at high frequencies. If there is a feedback path through a common power supply connection, oscillation will occur. In the case of IC op-amps, which are physically small, the input, output and power supply terminals are close, creating the ideal conditions for undesired feedback. To make

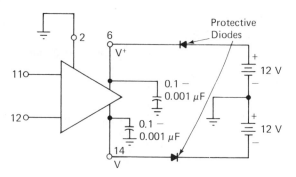

Fig. 2-2. Typical power supply connections for op-amp.

the problem worse, most op-amps are capable of passing frequencies higher than those specified on the datasheet.

For example, an op-amp to be used in the audio range (say up to 20 kHz with a power gain of 20 dB) could possibly pass a 10 MHz signal with some slight gain. This higher-frequency signal could be a harmonic of signals in the normal operating range and, with sufficient gain, could feed back and produce undesired oscillation.

When laying out any op-amps, particularly IC op-amps in the breadboard or experimental stage, always consider the circuit as being radio-frequency (RF), even though the op-amp is not supposed to be capable of RF operation, and the circuit is not normally used with RF.

2-1.3 Grounding metal IC op-amp cases

The metal case of the IC op-amp shown in Fig. 2-2 is connected to terminal 2 *and to no other point in the internal circuit.* Thus, terminal 2 can and should be connected to equipment ground, as well as to the common or ground of the two power supplies.

The metal cases of some IC op-amps may be connected to a point in the internal circuit. If so, the case will be at the same voltage as the point of contact. For example, the case might be connected to pin 14 of the op-amp shown in Figs. 2-1 and 2-2. If so, the case will be below ground (or "hot") by 12V. If the case is mounted directly on a metal chassis that is at ground, the op-amp and power supply will be damaged. Of course, not all IC op-amps have metal cases; likewise, not all metal cases are connected to the internal circuits. However, this point must be considered *before* using a particular IC op-amp.

2-1.4 Calculating current required for op-amps

The datasheets for op-amps usually specify a nominal operating voltage (and possibly a maximum operating voltage), as well as a "total device dissipation." These figures can be used to calculate the current required for a particular op-amp. Use simple dc Ohm's law and divide the power by the voltage to find the current. However, certain points must be considered.

First, use the actual voltage applied to the op-amp. The actual voltage should be equal to the nominal operating voltage, but in no event higher than the maximum voltage.

Second, use the *total* device dissipation. The datasheet may also list other power dissipations, such as "device dissipation," which is defined as the dc power dissipated by the op-amp itself (with output at zero and no load). The other dissipation figures will always be smaller than the total power dissipation.

2-1.5 Power supply tolerances

Typically, op-amps will operate satisfactorily with ±20 percent power supplies. These tolerances apply to actual operating voltage, not to maximum voltage limits. The currents (or power consumed) will vary proportionately.

Power supply ripple and regulation are both important. Generally, solid-state power supplies with filtering and full feedback regulation are recommended, particularly for high gain op-amps. Ideally, ripple (and all other noise) should be 1 percent or less.

The fact that an op-amp generally requires two power supplies (due to the differential amplifiers) creates a particular problem in regard to offset. For example, if the $V+$ supply is 20 percent high, and the $V-$ supply is 20 percent low, there will be an unbalance and offset, even though the op-amp circuits are perfectly balanced.

The effects of operating an op-amp beyond the voltage tolerance are essentially the same as those experienced when the op-amp is operated at temperature extremes. That is, a high power supply voltage will cause the op-amp to "overperform," whereas low voltages will result in "underperformance." A low voltage will usually not result in damage to the op-amp, as is the case when operating the op-amp beyond the maximum rated voltage.

2-1.6 Single power supply operation

An op-amp is generally designed to operate from symmetrical positive and negative power supply voltages. This results in a high common-mode rejection capability, as well as good low-frequency operation (typically a few Hz down to direct current). If the loss of very low frequency operation can be tolerated, it is possible to operate op-amps from a single power supply, even though designed for dual supplies. Except for the low frequency loss, the other op-amp characteristics should be unaffected.

The following notes describe a technique that can be used with most op-amps to permit operation from a single power supply, with a minimum of design compromise. The same maximum ratings that appear on the datasheet are applicable to the op-amp when operating from a single polarity power supply, and must be observed for normal operation. Likewise, all of the considerations discussed thus far in this chapter apply to single supply operation.

The technique described here is generally referred to as the "split Zener" method. The main concern in setting up for single supply operation is to *maintain the relative voltage levels.* With an op-amp designed for dual supply operation, there are three reference levels: $+V$, 0 and $-V$. For example, if the datasheet calls for $\pm 12V$ supplies, the three reference levels are: $+12V$, $0V$ and $-12V$.

For single supply operation, these same reference levels can be maintained by using $++V$, $+V$ and ground (that is, $+24V$, $+12V$ and $0V$), where $++V$ represents a voltage level double that of $+V$. This is illustrated in Fig. 2-3 where the op-amp is connected in the split-Zener mode. Note that there is no change in the *relative voltage levels* even though the various op-amp terminals are at different voltage levels (with refer-

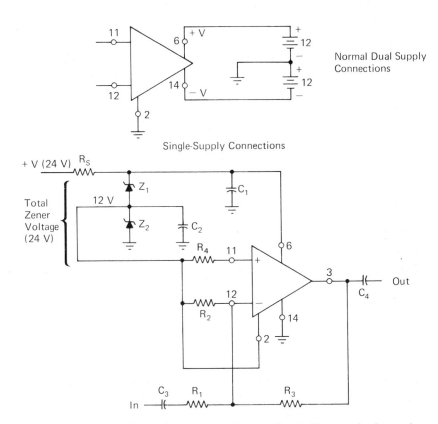

Fig. 2-3. Connections for single power supply operation (with ground reference).

ence to ground). Terminal 14 (normally connected to the −12V supply) is at ground. Terminal 2 (normally ground or common) is set at one-half the total Zener voltage (+12V). Terminal 6 (normally connected to the +12 supply) is set at the full Zener voltage (+24V).

With single supply, the differential input terminals (11 and 12), which are normally at ground in a dual supply system, must also be raised up one-half the Zener voltage (+12V). Under these circumstances, the output terminal (3) will also be at one-half the Zener voltage, plus or minus any offset (refer to Sec. 1-3.8 of Chapter 1).

To minimize offset errors due to unequal voltage drops caused by the input bias currents across unequal resistances, it is recommended that the value of the input offset resistance R_4 be equal to the parallel combination of R_2 and R_3 (as discussed in Sec. 1-3.8).

Any deviation between absolute Zener level will also contribute to an error in the op-amp output level. Typically, this is on the order of 50 to 100 μV per volt of deviation of Zener level. Except in rare cases, this deviation should be of little concern.

Note that the op-amp of Fig. 2-3 has a ground reference terminal (terminal 2). Not all op-amps have such terminals. Some op-amps have only $+V$ and $−V$ terminals or leads even though the two levels are referenced to a common ground. That is, there is no physical ground terminal or lead on the op-amp.

Figure 2-4 shows the split-Zener connections for single supply operation with such op-amps. Here, the input terminals (A and B) are set at one-half the total Zener supply voltage; the $−V$ terminals are set at ground and the $+V$ terminals are at the full Zener voltage (+24V).

Figures 2-3 and 2-4 show connection to positive power supplies. Negative power supplies can also be used. With a negative supply, the $+V$ terminal is connected to ground and the $−V$ terminal is connected to total Zener supply (+24V), with the input terminals and op-amp ground terminal (if any) connected to one-half the Zener supply. Of course, the polarity of the Zener diodes must be reversed.

Figures 2-3 and 2-4 both show a series resistance R_S for the Zener diodes. This is standard practice for Zener operation. The approximate or trial value for R_S is found by:

$$\frac{(\text{maximum supply voltage} - \text{total Zener voltage})^2}{\text{safe power dissipation of Zeners}}$$

For example, assume that the total Zener voltage is 24V (12V for each Zener), that the supply voltage may go as high as 27V, and that 2 watt Zeners are used. Under these conditions:

$$\frac{(27-24)^2}{2} = \frac{(3)^2}{2} = 4.5 \text{ ohm for } R_S$$

Effects of single supply operation. From a user's standpoint, operation of an op-amp with a single supply is essentially the same as with the conventional dual power supply. The following notes describe the basic differences in operational characteristics with both types of power supplies.

The normal op-amp phase/frequency compensation methods are the same for both types of supplies. The high-frequency limits are essentially the same. However, the low frequency limit of an op-amp with a single supply is set by the values of capacitors C_3 and C_4. These capacitors are not required for dual supply operation. Capacitors C_3 and C_4 are required for single supply operation since both the input and output of the op-amp are at a voltage level equal to one-half the total Zener voltage

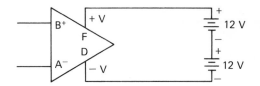

Normal Dual-Supply Connections

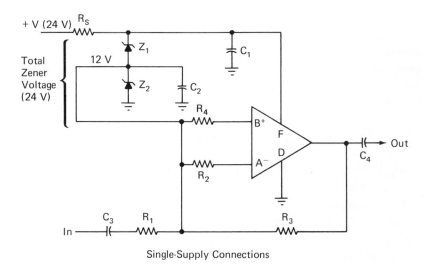

Single-Supply Connections

Fig. 2-4. Connections for single supply operation (without ground reference).

(or 12V using our example). Thus, the op-amp can not be used as a direct current amplifier with the single supply system. In a dual supply system the inputs and outputs are at 0V.

The closed-loop gain is the same for both types of supplies, and is determined by the ratio of R_3/R_1.

The values of decoupling capacitors C_1 and C_2 are essentially the same for both types of supplies. However, it may be necessary to use slightly larger values with the single supply system, since the impedance of the Zeners is probably different from that of the power supply (without Zeners).

The value of R_2 should be between 50K and 100K for a typical op-amp. Values of R_2 much higher or lower than these limits can result in decreased gain or in an abnormal frequency response. From a practical standpoint choose trial values using the guidelines and then run gain and frequency response tests.

The value of R_4, the input offset resistance, is chosen to minimize offset error from impedance unbalance. As an approximate trial value, the resistance of R_4 should be equal to the parallel combination of R_2 and R_3. That is,

$$R_4 \approx \frac{R_2 R_3}{R_2 + R_3}$$

2-2. BASIC OP-AMP SYSTEM DESIGN

Figure 2-5 is the working schematic of a closed-loop op-amp system, complete with external circuit components. The design considerations discussed in Chapter 1 and in Sec. 2-1 apply to the circuit of Fig. 2-5. The following paragraphs provide a specific design example for the circuit.

2-2.1 Op-amp characteristics

Supply voltage: +15 and −15V nominal, ±19V maximum

Total device dissipation: 750 mW, derating 8 mW/°C

Temperature range: 0°C to +70°C

Input offset voltage: 3 mV typical

Input offset current: 10 nA nominal, 30 nA maximum

Input bias current: 100 nA nominal, 200 nA maximum

Input offset voltage sensitivity: 0.2 mV/V

Device dissipation: 300 mW maximum

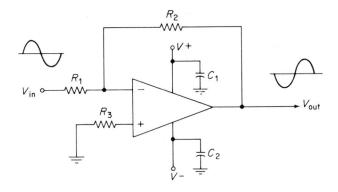

$$\text{Voltage gain} = \frac{V_{out}}{V_{in}} = \frac{R_2}{R_1} \qquad C_1 = C_2 = 0.1 - 0.001 \,\mu F$$

$$R_3 = \frac{R_1 R_2}{R_1 + R_2}$$

Fig. 2-5. Basic op-amp system connections.

Open-loop voltage gain: as shown in Fig. 1-22

Slew rate: 4 V/μS at a gain of 1, 6V/μS at a gain of 10, 33V/μS at a gain of 100

Open-loop bandwidth: as shown in Fig. 1-22

Common-mode rejection: 94 dB

Output voltage swing: 23V (p-p) typical

Input impedance: 1 megohm

Output impedance: 300 ohms

Input voltage range: −13V, +10V

Output power: 250 mW typical

2-2.2 Design example

Assume that the circuit of Fig. 2-5 is to provide a voltage gain of 100 (40 dB), the input signal is 80 mV(RMS), the input source impedance is not specified, the output load impedance is 500 ohms, the ambient temperature is 25°C, the frequency range is dc up to 300 kHz, and the power supply is subject to 10 percent variation.

Frequency/gain relationship. Before attempting to calculate any circuit values, make certain that the op-amp can produce the desired voltage gain at the maximum frequency. This can be done by reference to a graph similar to Fig. 1-22.

Note that the maximum frequency of 300 kHz intersects the phase curve at about 135°. This allows a phase margin of 45°. Thus, the op-amp should be stable over the desired frequency range.

Note that the maximum frequency of 300 kHz intersects several capacitance curves above the 40 dB open-loop voltage gain level. Thus, the op-amp should be able to produce more than 40 dB of open-loop gain.

Supply voltage. The positive and negative supply voltages should both be 15V since this is the nominal value listed. Most op-amp datasheets will list certain characteristics as "maximum" (temperature range, total dissipation, maximum supply voltage, maximum input signal, etc.), and then list the remaining characteristics as "typical" with a given "nominal" supply voltage.

In no event can the supply voltage exceed the 19V maximum. Since the available supply voltage is subject to a 20 percent variation, or 18V maximum, the supply is within the 19V limit.

Decoupling or bypass capacitors. The values of C_1 and C_2 should be found on the datasheet. In the absence of a value, use 0.1 μF for any frequency up to 10 MHz. If this value produces a response problem at any frequency (high or low), try a value between 0.001 and 0.1 μF.

Closed-loop resistances. The value of R_2 should be 100 times the value of R_1 to obtain the desired gain of 100. The value of R_1 should be selected so that the voltage drop across R_1 (with the nominal input bias current) is comparable to the input signal (never larger than the input signal).

A 50 ohm value for R_1 will produce a 10 μV drop with the maximum 200 nA input bias current. Such a 10 μV drop is less than 10 percent of the 80 mV input signal. Thus, the fixed drop across R_1 should have no appreciable effect on the input signal.

With a 50 ohm value for R_1, the value of R_2 must be 5K (50 × 100 gain = 5000).

Offset minimizing resistance. The value of R_3 can be found using the equation of Fig. 2-5 once the values of R_1 and R_2 have been established. Note that the value of R_3 works out to about 49 ohms, using the Fig. 2-5 equation:

$$\frac{R_1 R_2}{R_1 + R_2} = \frac{50 \times 5000}{50 + 5000} \approx 49 \text{ ohms}$$

Thus, a simple trial value for R_3 is always *slightly less* than the R_1 value. The final value of R_3 should be such that the no-signal voltages at *each input are equal.*

Comparison of circuit characteristics. Once the values of the external circuit components have been selected, the characteristics of the op-amp and the closed-loop circuit should be checked against the requirements of the design example. The following is a summary of the comparison.

Gain versus phase compensation. The closed-loop gain should always be less than the open-loop gain. As a guideline, the open-loop gain should be at least 20 dB greater than closed-loop gain. Figure 1-22 shows that open-loop gain up to about 66 dB is possible with a proper phase compensation capacitor. Figure 1-22 also shows that a capacitance of 0.001 μF (1000 pF) will produce an open-loop gain of slightly less than 60 dB at 300 kHz, whereas a 300 pF capacitance will produce a 66 dB open-loop gain. To assure an open-loop gain of 60 dB, use a capacitance value of about 700 pF.

With a 60 dB open loop gain, and values of 50 and 5000 ohms, respectively, for R_1 and R_2, the closed-loop circuit should have a flat frequency response of 40 dB (gain of 100) from zero up to 300 kHz. A rolloff will start at frequencies above 300. Thus, the closed-loop gain is well within tolerance.

Input voltage. The *peak input voltage* must not exceed the rated maximum input signal. In this case, the rated maximum is +10V and −13V, whereas the input signal is 80 mV (RMS), or approximately 112 mV peak (80 × 1.4). This is well below the +10V maximum limit.

When the rated maximum input signal is an uneven positive and negative value, always use the *lowest value* for total swing of the input signal. In this case, the input swings from +112 mV to −112 mV, far below +10V.

An input signal that started from zero could swing as much as +10V and −10V, without damaging the op-amp. An input signal that started from −2V could swing as much as ±11V.

Output voltage. The *peak-to-peak output voltage* must not exceed the rated maximum output voltage swing (with the required input signal and selected amount of gain).

In this case, the rated output voltage swing is 23V (p-p), whereas the actual output is approximately 22.4V (80 mV RMS input × a gain of 100 = 8000 mV output; 8000 mV × 2.8 = 22.4V peak-to-peak). Thus, the anticipated output is within the rated maximum.

However, the actual output voltage is dependent upon slew rate, which, in turn, depends upon compensation capacitance. As shown in the characteristics, the slew rate is given as 33 V/μS when gain is 100. The datasheet does not show a relationship between slew rate and compensation capacitance. However, since slew rate is always maximum with the lowest value of compensation capacitance, it can be assumed that the slew rate will be 33 V/μS with a capacitance of 700 pF (which is near the lowest recommended value of 300 pF, Fig. 1-22).

Using the equations in Sec. 1-3.3, it is possible to calculate the output voltage capability of the op-amp. With a maximum operating frequency of 300 kHz, and an assumed slew rate of 33 V/μS, the peak output voltage capability is:

$$\frac{33,000,000}{6.28 \times 300,000} \approx 17V$$

Thus, the peak-to-peak output voltage capability is 34V, well above the anticipated 22.4V. A quick approximation of the output voltage capability can also be found by reference to Fig. 1-39.

Output power. The output power of an op-amp is usually computed on the basis of RMS output voltage (rather than peak or peak-to-peak) and output load.

In this example, the output voltage is 8V RMS (80 mV $\times$ 100 gain 8000 mV = 8V). The load resistance or impedance is 500 ohms as stated in the design assumptions. Thus, the output power is:

$$\frac{(8)^2}{500} = 0.128 \text{ W} = 128 \text{ mW}$$

A 128 mW output is well below the 250 mW typical output power of the op-amp. Also, 128 mW output plus a device dissipation of 300 mW is 428 mW, well below the rated 750 mW total device dissipation. Thus, the op-amp should be capable of delivering full power output to the load.

Note that power output ratings usually are at some given temperature — 25°C in this case. Assume that the temperature is raised to 50°C. The total device dissipation must be derated by 8 mW/°C, or a total of 200 mW for the 25°C increase in temperature. This derates or reduces the 750 mW total device dissipation to 550 mW. However, the 550 mW is still well above the anticipated 428 mW.

Output impedance. Ideally, the closed-loop output impedance should be as low as possible and always less than the load impedance. The closed-loop output impedance can be found using the equations of Fig. 1-16. In this example, the approximate output impedance will be:

$$\frac{300}{1 + 1000 \times \left(\dfrac{50}{50 + 5000}\right)} \approx 30 \text{ ohms}$$

2-3. ZERO OFFSET SUPPRESSION

Figure 2-6 is the working schematic of an op-amp used to provide amplification of a small signal riding on a large, fixed direct current level. For example, assume that the input signal varies between 3 and 8 mV, and that the signal source never drops below 7V. That is, the source is +7V with no signal, and +7.003 to +7.008V with signal. Now assume that the output is to vary between 300 and 800 mV.

The obvious solution is to apply a fixed +7V to the noninverting input. This will offset the +7V at the inverting input, and result in 0V output (under no-signal conditions). This solution ignores the fact that the op-amp probably has some input offset, or assumes that the op-amp has some provisions for neutralizing the offset. The solution also assumes that the signal is riding on exactly +7V.

If input offset cannot be ignored (say because it is large in relation to the signal), or if the fixed dc voltage is subject to possible change, the alternate offset circuit of Fig. 2-6 should be used. The circuit in Fig. 2-6 is a simple resistance network that makes use of existing V_{CC} and V_{EE} voltages. In use, potentiometer R_4 is adjusted to provide zero output from the op-amp under no-signal conditions. That is, the 3 to 8 mV signal is removed, but the +7V remains at the inverting input, while R_4 is adjusted for zero at the op-amp input.

The values for the offset network are not critical. However, the values

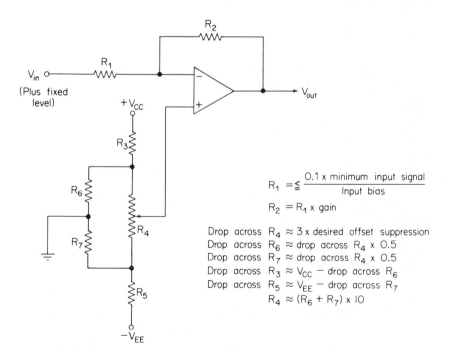

$$R_1 = \leq \frac{0.1 \times \text{minimum input signal}}{\text{Input bias}}$$

$$R_2 = R_1 \times \text{gain}$$

Drop across $R_4 \approx 3 \times$ desired offset suppression
Drop across $R_6 \approx$ drop across $R_4 \times 0.5$
Drop across $R_7 \approx$ drop across $R_4 \times 0.5$
Drop across $R_3 \approx V_{CC} -$ drop across R_6
Drop across $R_5 \approx V_{EE} -$ drop across R_7
$$R_4 \approx (R_6 + R_7) \times 10$$

Fig. 2-6. Op-amp with zero offset suppression to provide amplification of small signal riding on large, fixed level.

should be selected so that a minimum of current is drawn from the $V_{CC} - V_{EE}$ supplies, and a minimum of current should flow through R_4. This is discussed further in the design example.

The technique shown in Fig. 2-6 can be applied to most of the op-amp applications described in this chapter and Chapter 4. That is, any zero offset (at the op-amp input and output) can be suppressed by applying a fixed (or adjustable) voltage of correct polarity and amplitude to the opposite input.

For example, in the basic circuit in Fig. 2-5 (Sec. 2-2) the offset can be suppressed by application of a positive voltage at the noninverting input, in place of R_3 (or in addition to R_3). The offset suppression voltage can be fixed, or adjusted, using the circuits of Fig. 2-6 as applicable.

2-3.1 Design example

Assume that the circuit in Fig. 2-6 is used to monitor a 3 to 8 mV signal, riding on a +7V level, and to produce a 300 mV to 800 mV output (a gain of 100). The op-amp does not have provisions for input offset neutralization or null. It is important that the no-signal output be exactly 0V. The op-amp has an input bias of 200 nA, and V_{CC} and V_{EE} are 15V. The voltage drop across R_1 due to the input bias current should be no greater than 10 percent of the lowest input signal. With a low signal of 3 mV, and a 200 nA bias current, the value of R_1 is 1500 ohms (3 mV $\times$ 0.1 = 0.3 mV; 0.3 mV/200 nA = 1500). With R_1 at 1500, and a required gain of 100, the value of R_2 is 150K.

With V_{CC} and V_{EE} at +15V and −15V, respectively, the total drop across the offset adjustment network is 30V. Allowing an arbitrary 1 mA current through the network, the total resistance should be about 30K. Since the desired offset suppression is approximately 7V (to offset the +7V level), the drop across R_4 should be approximately 21V. This results in a drop of about 10.5V each across R_6 and R_7. In turn, the drops across R_3 and R_5 (each) should be about 4.5V(15V −10.5V). With a desired 4.5V drop, and approximately 1 mA current flow, the values of R_3 and R_5 should be 4500 ohms each. Likewise, the values of R_6 and R_7 should be 10.5K each. With R_6 and R_7 at 10.5K each, the value of R_4 should be 220K ($R_6 + R_7$ = 22K; 22K $\times$ 10 = 220K).

2-4. VOLTAGE FOLLOWER (SOURCE FOLLOWER)

Figure 2-7 is the working schematic of an op-amp used as a voltage follower (also known as a source follower). The circuit is essentially a *unity gain amplifier*. There is no feedback or input resistance in the circuit. Instead, the output is fed back directly to the inverting input. Signal input is applied directly to the noninverting input. With this arrangement, the output voltage equals the input voltage (or may be slightly less than the input voltage). However, the input impedance is very high, with the output impedance very low (as shown by the equations).

In effect, the input impedance of the op-amp is multiplied by the open-loop gain, whereas the output impedance is divided by the open-loop gain. Keep in mind that open-loop gain varies with frequency. Thus, the closed-loop impedances are frequency dependent.

Another consideration that is sometimes overlooked in this application is the necessity of supplying the input bias current to the op-amp. In a conventional circuit, bias is supplied through the input resistances. In the circuit of Fig. 2-7, the input bias must be supplied through the signal source. This may alter input impedance.

One more problem with the circuit in Fig. 2-7 is that the total input voltage is a common-mode voltage. That is, a voltage equal to the total input voltage appears across the two inputs. Should the input signal consist of a large dc value, plus a large signal variation, the common-mode input voltage range may be exceeded. One solution for the problem is to capacitively couple the input signal to the noninverting input. This eliminates any dc voltage, and only the signal appears at the inputs. This will solve the common-mode problem, but will require the use of a resistor from the noninverting input to ground. The resistance provides a path for the input bias current. However, the resistance also sets the input impedance of the op-amp. If the resistance is large, the input bias current will produce a large offset voltage (both input and output) across the resistance.

Some manufacturers recommend that resistances (of equal value) be

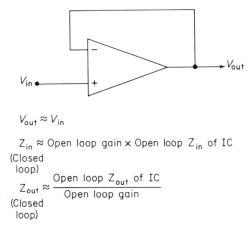

$$V_{out} \approx V_{in}$$

$Z_{in} \approx$ Open loop gain $\times$ Open loop Z_{in} of IC
(Closed
loop)

$$Z_{out} \approx \frac{\text{Open loop } Z_{out} \text{ of IC}}{\text{Open loop gain}}$$

(Closed
loop)

Fig. 2-7. Voltage follower (unity gain) using an op-amp.

used in the feedback loop (from output to inverting input) and at the noninverting input. This will still result in unity gain, and the circuit will perform as a voltage or source follower. However, the input and output impedances will then be set (primarily) by the resistance values rather than the op-amp characteristics, as is the case with the circuit in Fig. 2-7.

Another recommendation is that the circuit in Fig. 2-7 be followed by a power amplifier capable of high current output. This is shown in Fig. 2-8. Here, the op-amp output is fed to the power amplifier input. The output from the power amplifier is fed back to the op-amp inverting input. The voltage gain remains at approximately 1, but the available current is set by the power amplifier capability. The input and output impedances of the overall circuit are increased by the gains of both amplifiers. For example, if the op-amp and power amplifier open-loop gains are both 100, the input impedance is multiplied by approximately 10,000 and the output impedance is divided by 10,000.

2-4.1 Design example

Assume that the circuit in Fig. 2-7 is to provide unity gain with high input impedance and low output impedance. Also assume that the op-amp has an open-loop gain of 100 (40 dB) at a specific frequency, an output impedance of 300 ohms and an input impedance of 1 megohm. With these characteristics, the closed-loop input impedance is: 100×1 megohm ≈ 100 megohms. The closed-loop output impedance is: $300/100 \approx 3$ ohms.

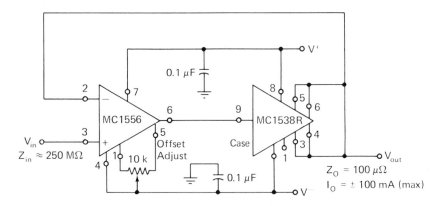

Fig. 2-8. High input impedance, high output current voltage follower (Courtesy Motorola).

2-5. UNITY GAIN WITH FAST RESPONSE

One of the problems of a unity gain amplifier is that the slew rate is very poor. That is, the response time is very slow, and the power bandwidth is decreased. The reason for poor bandwidth with unity gain is that most op-amp datasheets recommend a large-value compensating capacitor for unity gain.

As an example, assume that the op-amp has the characteristics as shown in Fig. 1-22, and the desired operating frequency is 200 kHz (the unity gain op-amp must have a full power bandwidth up to 200 kHz). The recommended compensation for 60 dB gain is 0.001 μF, whereas the unity gain compensation is 1 μF. Now assume that the op-amp also has the characteristics of Fig. 1-40, and that the load is 100 ohms. With a compensation of 0.001 μF, full power can be delivered at frequencies well above 200 kHz. However, if the compensating capacitor is 1 μF, the maximum frequency at which full power can be delivered is below 4 kHz.

Several methods are used to provide fast response time (high slew rate) and a good power bandwidth with unity gain. Two such methods are described next.

2-5.1 Design examples

Using op-amp datasheet phase compensation. The circuit in Fig. 2-9 shows a method of connecting an op-amp for unity gain, but with high slew rate (fast response and good power bandwidth). With this circuit, the phase compensation recommended on the datasheet is used, but with modifications. Instead of using the unity gain compensation, use the datasheet phase compensation recommended for a gain of 100. Then select values of R_1 and R_3 to provide unity gain ($R_1 = R_3$). As shown by the equations, the values of R_1 and R_3 must be approximately 100 times the value of R_2. Thus, R_1 and R_3 must be fairly high values for practical design.

Assume that the circuit in Fig. 2-9 provides unity gain, but with a slew rate approximately equal to that which results when a gain of 100 is used. Also assume that the input/output signal is 8V and that the op-amp has an input bias current of 200 nA and compensation characteristics similar to those of Fig. 1-22, but with higher gain.

The compensation capacitance recommended for a gain of 100 is 0.001 μF. With this established, select the values of R_1, R_2 and R_3. The value of R_1 (and, consequently, that of R_3) is selected so that the voltage

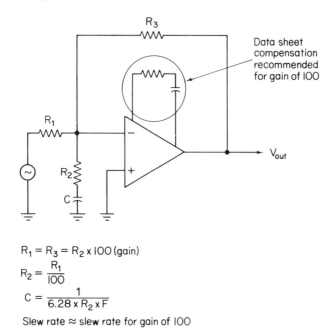

$$R_1 = R_3 = R_2 \times 100 \text{ (gain)}$$

$$R_2 = \frac{R_1}{100}$$

$$C = \frac{1}{6.28 \times R_2 \times F}$$

Slew rate ≈ slew rate for gain of 100

Fig. 2-9. Unity gain op-amp with fast response (good slew rate) using datasheet phase compensation.

drop with nominal input bias current is comparable (preferably 10 percent or less) to the input signal. Ten percent of the 8V input is 0.8V. Using the 0.8V value and the 200 nA input bias, the resistance of R_1 is 4 megohms. With R_1 at 4 megohms, R_3 must also be 4 megohms, and R_2 must be 40K (4 megohms/100 = 40K).

The value of C_1 is found using the equations in Fig. 2-9 once the value of R_2 and the open-loop rolloff point are established. Assume that the op-amp has characteristics similar to those shown in Fig. 1-19, where the 6 dB/octave rolloff starts at about 200 kHz. With these figures, the value of C is:

$$C \approx \frac{1}{6.28 \times 40K \times 200 \text{ kHz}} \approx 19 \text{ pF}$$

Using input phase compensation. The circuit in Fig. 2-10 shows a method of connecting an op-amp for unity gain, but with high slew rate, using input compensation. With this circuit, the phase compensation recommended on the op-amp datasheet is not used. Instead, the input phase compensation system in Sec. 1-2.5 is used.

The first step is to compensate the op-amp by modifying the open-loop

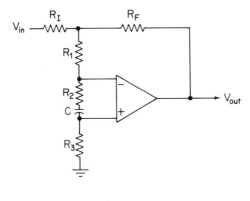

R₁,R₂,R₃,C = see text
$R_I = R_F = 0.25 \times R_1$

Fig. 2-10. Unity gain op-amp with fast response (good slew rate) using input phase compensation.

input impedance, as described in Sec. 1-2.5 and Fig. 1-20. The recommended values are: $R_1 = R_3 = 5K$, $R_2 = 2100$, $C = 0.0004$ μF.

Next, select values of R_I and R_F to provide unity gain (both R_I and R_F must be the same value). The values of R_I and R_F are not critical, but they must be identical. Using the equations of Fig. 2-10, the values of R_I and R_F should be 0.25 times the value of R_1, or $0.25 \times 5K\Omega = 1250\Omega$.

2-6. SUMMING AMPLIFIER (ANALOG ADDER)

Figure 2-11 is the working schematic of an op-amp used as a summing amplifier (or analog adder). When open loop gain is high, the circuit functions with a minimum of error to sum a number of voltages. One circuit input is provided for each voltage to be summed. The single circuit output is the sum of the various input voltages (a total of three in this case), multiplied by any circuit gain. Generally, gain is set so that the output is at some given voltage value when all inputs are at their maximum value. In other cases, the resistance values are selected for unity gain.

2-6.1 Design example

Assume that the circuit of Fig. 2-11 is to be used as a summing amplifier to sum three input voltages. Each of the voltage inputs varies from 8 to 30 mV (RMS). The output must be at least 1V(RMS)

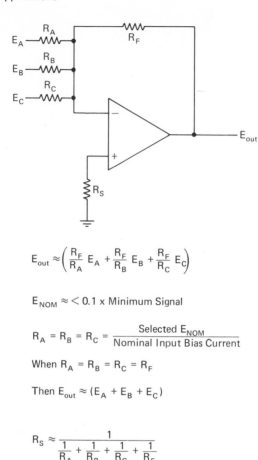

$$E_{out} \approx \left(\frac{R_F}{R_A} E_A + \frac{R_F}{R_B} E_B + \frac{R_F}{R_C} E_C \right)$$

$$E_{NOM} \approx\, < 0.1 \times \text{Minimum Signal}$$

$$R_A = R_B = R_C = \frac{\text{Selected } E_{NOM}}{\text{Nominal Input Bias Current}}$$

When $R_A = R_B = R_C = R_F$

Then $E_{out} \approx (E_A + E_B + E_C)$

$$R_S \approx \frac{1}{\dfrac{1}{R_A} + \dfrac{1}{R_B} + \dfrac{1}{R_C} + \dfrac{1}{R_F}}$$

Fig. 2-11. Summing amplifier (analog adder).

with full input on all three channels, but must not exceed 2V (RMS) at any time.

A summing amplifier application will generally require some phase compensation, if accuracy is a major concern. In selecting phase compensation, always use the values recommended on the datasheet for the highest gain condition. As shown by the equations, phase compensation must be suitable for the closed loop gain ratio of R_F to the parallel equivalent of R_A, R_B and R_C. Do not use the input impedance modification compensation described in Sec. 1-2.5.

To simplify design, make resistors R_A, R_B and R_C the same value. Note that the input bias current is then divided equally, and produces the same voltage drop across each resistor.

Select the values of R_A, R_B and R_C so that the voltage drop across them (with nominal input bias current) is less than 10 percent of the minimum input signal. Assume a 200 nA input bias current. With three equal resistances, this results in about 66 nA through each resistor. With a minimum input signal of 8 mV, the desired maximum drop is 0.8 mV. With a drop of 0.8 mV, and an approximate 66 nA current flow, the maximum values for R_A, R_B and R_C are 12K(0.8mV/66 nA). A 10K value will produce an approximate 0.66 mV drop, which is less than 10 percent of the 8 mV minimum input signal.

The total (or maximum possible) signal voltage at the input is 90 mV (3 × 30 mV). Thus, the value of R_F should be between 11 and 22 times that of R_A through R_C to get a minimum of 1V and a maximum of 2V (1V/90 mV ≈ 11; 2V/90 mV ≈ 22).

With a 10K value for R_A through R_C, the value of R_F could be 110K to 220K. Assume that the 220K value is selected. Under these conditions, the circuit output is 22 times the sum of the three input voltages or 528 mV (all three inputs minimum) to 2V (all three inputs maximum).

The value of the offset minimizing resistance R_S is found using the equation of Fig. 2-11 once the values of R_A, R_B, R_C and R_F have been established:

$$R_S = \frac{1}{\dfrac{1}{10,000} + \dfrac{1}{10,000} + \dfrac{1}{10,000} + \dfrac{1}{220,000}} \approx 3238 \text{ ohms}$$

2-7. SCALING AMPLIFIER (WEIGHTED ADDER)

Figure 2-12 is the working schematic of an op-amp used as a scaling amplifier or weighted adder. The circuit is essentially the same as a summing amplifier, except that in the former the inputs to be summed are "weighted" or compensated to produce a given output range, or a given relationship between inputs. The weighting operation is possible because the condition that exists at the junction of the feedback resistor and the inverting input isolates each signal channel from the others. Each input signal enters through an impedance of such value that the ratio to the feedback impedance is equal to the desired weighting factor.

For example, assume that there are two inputs to be summed and that one input has a nominal voltage range eight times that of the other. Now assume that the output voltage range must be the same for both inputs. This can be done by making the input resistance for the low voltage input one-eighth the value of the high-input resistance. Since the feedback resistance is the same for both inputs, and gain (or output) is set by the

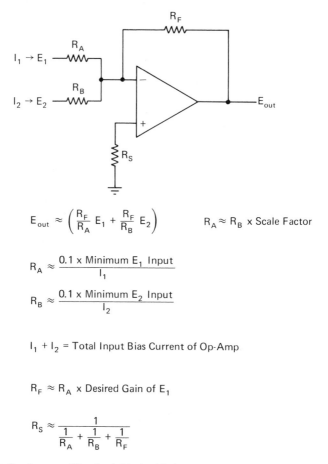

$$E_{out} \approx \left(\frac{R_F}{R_A} E_1 + \frac{R_F}{R_B} E_2 \right)$$ $R_A \approx R_B$ x Scale Factor

$$R_A \approx \frac{0.1 \text{ x Minimum } E_1 \text{ Input}}{I_1}$$

$$R_B \approx \frac{0.1 \text{ x Minimum } E_2 \text{ Input}}{I_2}$$

$I_1 + I_2 = $ Total Input Bias Current of Op-Amp

$R_F \approx R_A$ x Desired Gain of E_1

$$R_S \approx \frac{1}{\frac{1}{R_A} + \frac{1}{R_B} + \frac{1}{R_F}}$$

Fig. 2-12. Scaling amplifier (weighted adder).

ratio of feedback-to-input resistance, the low input is multiplied eight times as much as the high input. Thus, the output range is the same for both inputs.

In other cases, the circuit is used to make two equal inputs produce two outputs of different voltage ranges. Likewise, the circuit can be weighted so that two unequal inputs are made more (or less) unequal by a given *scale factor*. For example, if the inputs are normally 8 to 1, they can be made 3 to 1, or 7 to 1 or any practical value within the limits of the op-amp.

The minimum phase compensation needed for the circuit is that required for the gain obtained when a single signal drives all of the input channels in parallel. For example, assume that both channels are driven by a 1 mV in-phase signal, that one channel has a gain of 8 and the other channel has a gain of 3. The output will then be 11 mV (3 + 8). Under these conditions, use a datasheet phase compensation recommended for a gain of 11 (or the nearest gain value). Do not use the input impedance modification compensation described in Sec. 1-2.5.

2-7.1 Design example

Assume that the circuit in Fig. 2-12 is to be used as a scaling amplifier (weighted adder) for two inputs. One input has a nominal voltage range of 2 to 50 mV. The other input has a higher voltage range, 4 to 100 mV. The output range is to be the same for both inputs. That is, a full swing (2 to 50 mV) of the low input must produce the same output swing as a full swing of the high input (4 to 100 mV). In no case can the output exceed 2V.

Since R_F is common to both R_A and R_B, the value of R_B must be twice that of R_A, if the output is to be equal (under maximum and minimum input conditions). The input bias current will divide unequally between the two input resistors, with the current through R_A twice that of R_B. In this way, the fixed (no-signal) voltage drop produced by the bias is the same for both R_A and R_B.

Select the values of R_A and R_B so that the voltage drop across them (with nominal input bias current) is 10 percent of the minimum input signal. Assume a 6000 nA input bias current.

Since the current through R_A is twice that of R_B, the R_A current is two-thirds of the total, or 4000 nA, whereas the R_B current is 2000 nA.

Since the lowest input voltage is 2 mV, design starts with the low input resistor R_A. Ten percent of 2 mV is 0.2 mV. Using this 0.2 mV value and a 4000 nA input bias, the resistance of R_A is 50 ohms. With R_A at 50 ohms, R_B is 100 ohms.

The total (or maximum possible) signal voltage at the op-amp input is 150 mV (50 mV + 100 mV). However, since R_B is twice the value of R_A (with R_F fixed), the gain for the low input is twice that of the high input. Thus, the maximum 50 mV input has the same effect as a 100 mV input. The effective maximum input at the op-amp is 200 mV (100 mV effective low input plus 100 mV actual high input), with one-half the output voltage being supplied by each input.

The value of R_F should be such that the maximum output is 1V (one-half the desired 2V) with a maximum signal at either input. With an input at R_B of 100 mV and an R_B resistance of 100 ohms, the value of R_F must be 1000 (1V/100 mV = 10; 10 × 100 = 1000). This value of R_F checks with the R_A value to provide a gain of 20 for the low input (1000/50 = 20).

A gain of 20 also results in an output of 1V for the low input. Thus, the maximum output with both inputs at maximum is 2V.

The value of the offset minimizing resistance R_S is found using the equation of Fig. 2-12, once the values of R_A, R_B and R_F have been found:

$$R_S = \frac{1}{\frac{1}{50} + \frac{1}{1000} + \frac{1}{1000}} \approx 32 \text{ ohms}$$

2-8. DIFFERENCE AMPLIFIER
(ANALOG SUBTRACTOR)

Figure 2-13 is the working schematic of an op-amp used as a difference amplifier or analog subtractor. With either application, one signal voltage is subtracted from the other when signals are applied to both inputs simultaneously.

When the circuit is used as a *difference amplifier* the resistance values are chosen to provide a specific amount of gain. The gain is directly

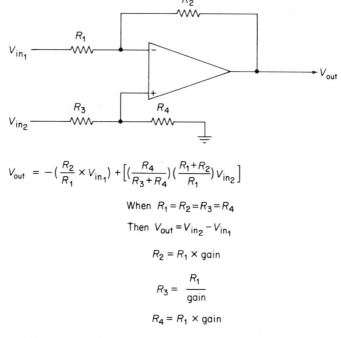

$$V_{out} = -\left(\frac{R_2}{R_1} \times V_{in_1}\right) + \left[\left(\frac{R_4}{R_3 + R_4}\right)\left(\frac{R_1 + R_2}{R_1}\right)V_{in_2}\right]$$

When $R_1 = R_2 = R_3 = R_4$

Then $V_{out} = V_{in_2} - V_{in_1}$

$R_2 = R_1 \times$ gain

$$R_3 = \frac{R_1}{\text{gain}}$$

$R_4 = R_1 \times$ gain

Fig. 2-13. Difference amplifier (analog subtractor).

proportional to the ratio of R_2/R_1. As a difference amplifier, the output is approximately equal to the algebraic sum (or difference) of the gains for the two input voltages, as shown by the equation of Fig. 2-13.

When the circuit is used as an *analog subtractor* the resistors are all made the same value. This provides no gain, but maximum accuracy. As an analog subtractor, the output is equal to the voltage at the non-inverting input, less the voltage at the inverting input. Of course, the output also represents the difference between the two input voltages, but without gain.

2-8.1 Design example.

Difference amplifier. When used as a difference amplifier, select the datasheet phase compensation recommended for the selected gain. Do not use the input impedance modification compensation described in Sec. 1-2.5.

Assume that the circuit in Fig. 2-13 is to be used as a difference amplifier. Both inputs vary from 8 to 70 mV. The output voltage is to be approximately the value of the V_2 input voltage, less the V_1 input voltage, multiplied by a factor of 30. Assume that the input bias current is 200 nA.

Ten percent of 8 mV is 0.8 mV. With 200 nA input bias, and a 0.8 mV (or less) drop, the maximum value of R_1 is 4000 ohms (0.8 mV/200 nA). Use 3000 ohms, which will produce a drop of 0.6 mV.

To provide a gain of 30, the value of R_2 must be 30 times the value of R_1, or $3000 \times 30 = 90K$. With a 3000 ohm value for R_1 and a gain of 30, the values of R_3 and R_4 are 100 and 90,000, respectively (using the equations of Fig. 2-13).

Now assume that the V_2 input is 50 mV and that the V_1 input is 40 mV. The output should be the difference (10 mV) multiplied by the gain (30), or 300 mV. However, using the equation of Fig. 2-13 and the values of R_1 through R_4 established in this example, note that the approximate output is slightly higher (about 333 mV). In practical applications, the two input voltages and the output voltages are measured. Then, if necessary, the value of R_3 is trimmed slightly to produce the precise difference voltage output. The same results can be obtained if the value of R_4 is trimmed, but it is usually more practical to work with R_3.

Analog subtractor. When used as an analog subtractor, select the datasheet phase compensation recommended for unity gain. Do not use the input impedance modification compensation described in Sec. 1-2.5.

Assume that the circuit of Fig. 2-13 is to be used as an analog subtractor. The inputs vary between 8 and 70 mV. The output must represent V_1 subtracted from V_2, and must not exceed 80 mV. Input bias current is 200 nA.

Since the output is to equal the value of V_2 less V_1, no gain is required. The value of 1 can be substituted for "gain" in all of the Fig. 2-13 equations. Thus, all resistance values are the same, and design can be based on a value for R_1. Since there is no gain, the output will never exceed 70 mV, which is below the required 80 mV maximum output.

Select a value for R_1 so that the voltage drop (with an input bias current of 200 nA) is 10 percent or less of the minimum input signal. Use 3000 ohms for R_1 (the same as the difference amplifier example). With R_1 at 3000 ohms, R_2, R_3 and R_4 must also be 3000 ohms each.

Now assume that the V_2 input is 50 mV and that the V_1 input is 40 mV (as in the difference amplifier example). Using the equation of Fig. 2-13, the output is exactly 10 mV. Of course, accuracy depends upon tolerances of the resistor values. Typically, R_1 through R_4 should be one percent, or better.

2-9. OP-AMP WITH HIGH INPUT IMPEDANCE

Figure 2-14 is the working schematic of an op-amp used as a high input impedance amplifier. The high input impedance and low output features of the unity gain amplifier (Sec. 2-4) are combined with modest gain, as shown by the equations of Fig. 2-14.

Note that the circuit of Fig. 2-14 is similar to that of the basic op-amp

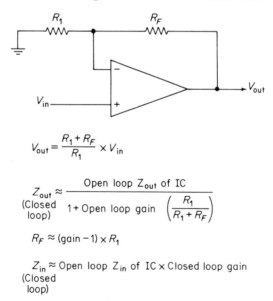

$$V_{out} = \frac{R_1 + R_F}{R_1} \times V_{in}$$

$$Z_{out} \approx \frac{\text{Open loop } Z_{out} \text{ of IC}}{1 + \text{Open loop gain} \left(\frac{R_1}{R_1 + R_F} \right)}$$
(Closed loop)

$$R_F \approx (\text{gain} - 1) \times R_1$$

$Z_{in} \approx$ Open loop Z_{in} of IC × Closed loop gain
(Closed loop)

Fig. 2-14. Op-amp with high input impedance.

(Sec. 2-2), except that there is no input offset compensating resistance (in series with the noninverting input) for the high input impedance circuit. This results in a tradeoff of higher input impedance, with some increase in output offset voltage.

In the basic op-amp (Sec. 2-2), an offset compensating resistance is used to nullify the input offset voltage of the op-amp. This (theoretically) results in no offset at the output. The output of the basic op-amp is at zero volts in spite of the tremendous gain.

In the unity gain amplifier (Sec. 2-4), there is no offset compensating resistance, but since there is no gain, the output is at the same offset as the input. In a typical IC op-amp, the input offset is less than 10 mV. This figure should not be critical for the output of a typical unity gain amplifier application.

In the circuit of Fig. 2-14, the offset compensation resistance is omitted. The output is offset by an amount equal to the input offset voltage of the op-amp, multiplied by the closed-loop gain. However, since the circuit of Fig. 2-14 is to be used for modest gains, modest output offset results.

2-9.1 Design example

Assume that the circuit in Fig. 2-14 is to provide a gain of 8, with maximum input impedance and minimum output impedance. Also assume that the op-amp has an open-loop gain of 1000 (60 dB), an output impedance of 300 ohms, an input impedance of 1M, and an input offset voltage of 3 mV. The minimum input signal is 20 mV. The maximum input signal is 100 mV. The input bias is 200 nA.

Select a value of R_1 so that the voltage drop is 10 percent (or less) of the minimum input signal, with nominal input bias current. Using the 20 mV minimum input signal and 200 nA input bias, the value of R_1 is 10K (20 mV $\times$ 0.1 = 2 mV; 2 mV/200 nA = 10,000 ohms).

With 10K for R_1 and a gain of 8, the value of R_F is:

$$R_F \approx (8 - 1) \times 10,000 \approx 70,000 \text{ ohms}$$

With a gain of 8, the closed-loop input impedance is:

$$Z_{in} \approx 1 \text{ M} \times 8 \approx 8 \text{ M}$$

The closed-loop output impedance is:

$$Z_{out} \approx \frac{300}{1 + 1000 \times \left(\dfrac{10,000}{10,000 \times 70,000}\right)} \approx 2.4 \text{ ohm}$$

With a gain of 8 and an output offset voltage of 3 mV, the output offset voltage is:

$$3 \text{ mV} \times 8 = 24 \text{ mV}.$$

2-10. NARROW BANDPASS AMPLIFIER (TUNED PEAKING)

Figure 2-15 is the working schematic of an op-amp used as a narrow bandpass amplifier (or tuned peaking amplifier).
Circuit gain is determined by the ratio of R_1 and R_F in the usual manner.

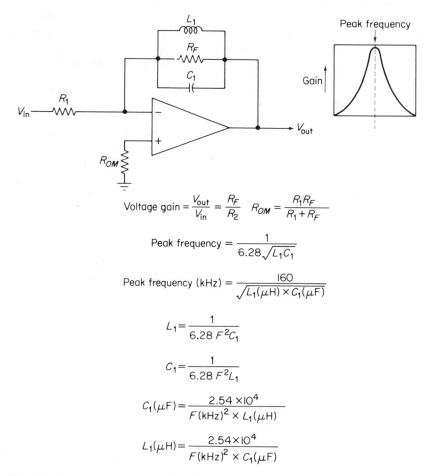

$$\text{Voltage gain} = \frac{V_{out}}{V_{in}} = \frac{R_F}{R_2} \qquad R_{OM} = \frac{R_1 R_F}{R_1 + R_F}$$

$$\text{Peak frequency} = \frac{1}{6.28\sqrt{L_1 C_1}}$$

$$\text{Peak frequency (kHz)} = \frac{160}{\sqrt{L_1(\mu H) \times C_1(\mu F)}}$$

$$L_1 = \frac{1}{6.28\, F^2 C_1}$$

$$C_1 = \frac{1}{6.28\, F^2 L_1}$$

$$C_1(\mu F) = \frac{2.54 \times 10^4}{F(\text{kHz})^2 \times L_1(\mu H)}$$

$$L_1(\mu H) = \frac{2.54 \times 10^4}{F(\text{kHz})^2 \times C_1(\mu F)}$$

Fig. 2-15. Narrow bandpass amplifier (tuned peaking).

However, the frequency at which maximum gain occurs (or the narrow band peak) is the resonant frequency of the L_1C_1 circuit. Capacitor C_1 and inductance L_1 form a parallel-resonant circuit that rejects the resonant frequency. Thus, there is a minimum feedback (and maximum gain) at the resonant frequency.

2-10.1 Design example

Assume that the circuit of Fig. 2-15 is to provide 40 dB gain at a peak frequency of 800 kHz. Use the datasheet phase compensation recommended for 40 dB, or the nearest gain value.

Select a value of R_1 on the basis of input bias current and voltage drop, as described for the closed loop resistances in Sec. 2-2.2. Assume an arbitrary value of 1K for R_1 to simplify this example. The value of R_{OM} is then the same, or slightly less.

With a value of 1K for R_1, the value of R_F is 100K (or $R_1 \times 100$) for a 40 dB gain.

Any combination of L_1 and C_1 can be used, provided the resonant frequency is 800 kHz. For frequencies below 1 MHz, the value of C_1 should be between 0.001 and 0.01 μF. Assume an arbitrary 0.002 μF for C_1. Using the equations in Fig. 2-15, the value of L_1 is:

$$L_1 = \frac{2.54 \times 10^4}{(800)^2 \times 0.002} \approx 20 \ \mu\text{H}$$

2-11. WIDE BANDPASS AMPLIFIER

Figure 2-16 is the working schematic of an op-amp used as a wide bandpass amplifier.

Maximum circuit gain is determined by the ratio of R_R and R_F. The gain of the passband or flat portion of the response curve is set by R_F/R_R. Minimum circuit gain is determined by the ratio of R_1 and R_F.

The ratios or relationships of R_F, R_R and R_N also determine the relationships of the passband frequencies. For example, if the value of R_F is increased in relationship to R_R, and all other factors remain the same, the frequency F_1 will be decreased, but with F_2 unchanged. Thus, there will be a greater frequency spread between F_1 and R_2.

The ratios given in the equations of Fig. 2-16 are selected to provide a differential of about 10 dB between minimum and maximum gain. That is, if minimum gain is 20 dB, maximum gain will be about 30 dB, and so on. In turn, minimum gain can be set by the ratio of R_1 to R_F.

There is also a direct relationship between the values of the capacitors, and between the capacitors and resistors. For example, if the value of C_N is increased in relation to C_R, and all other factors remain the same, the frequencies F_3 and F_4 will be decreased, with F_1 and F_2 unchanged. Thus, there will be a narrower passband. The same conditions occur for R_N (an increase in R_N decreases the passband).

It is obvious from this analysis of the equations that the shape of the passband is determined by capacitance and resistance ratios. Also, there must be a tradeoff between gain and frequency relationships. If the ratios

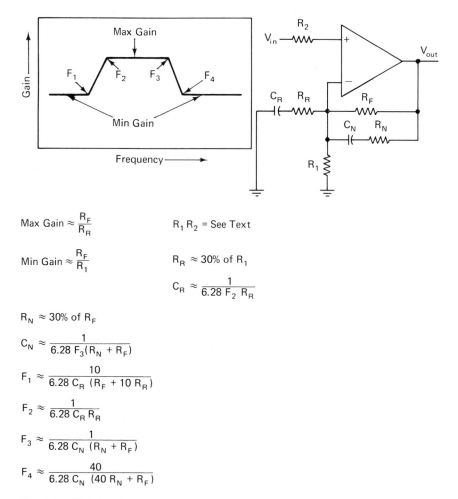

$$\text{Max Gain} \approx \frac{R_F}{R_R}$$

$$R_1 R_2 = \text{See Text}$$

$$\text{Min Gain} \approx \frac{R_F}{R_1}$$

$$R_R \approx 30\% \text{ of } R_1$$

$$C_R \approx \frac{1}{6.28 \, F_2 \, R_R}$$

$$R_N \approx 30\% \text{ of } R_F$$

$$C_N \approx \frac{1}{6.28 \, F_3 (R_N + R_F)}$$

$$F_1 \approx \frac{10}{6.28 \, C_R \, (R_F + 10 \, R_R)}$$

$$F_2 \approx \frac{1}{6.28 \, C_R \, R_R}$$

$$F_3 \approx \frac{1}{6.28 \, C_N \, (R_N + R_F)}$$

$$F_4 \approx \frac{40}{6.28 \, C_N \, (40 \, R_N + R_F)}$$

Fig. 2-16. Wide bandpass amplifier.

given in the equations of Fig. 2-16 do not provide the desired gain-frequency tradeoff, alter the ratios as needed.

2-11.1 Design example

Assume that the circuit of Fig. 2-16 is to provide approximately 100 dB minimum gain at all frequencies, and approximately 110 dB gain at the passband. Frequency F_2 is to be about 200 kHz, with frequency F_3 at about 400 kHz.

Use the phase compensation recommended for the 100 dB minimum gain, and not the passband gain of 110 dB.

Select a value of R_1 on the basis of input bias current and voltage drop, as described for the closed loop resistances in Sec. 2-2.2. Assume an arbitrary value of 1K for R_1 to simplify this example. The value of R_2 is then the same, or slightly less.

With a value of 1K for R_1, a value of 100K is used for R_F. With a value of 100K for R_F, the value of R_N is about 30K. Also, with R_1 at 1K, R_R is 300 ohms. These relationships produce a minimum gain of 100 dB and a maximum gain of 110 dB.

With a value of 300 ohms for R_R, and an F_2 of 200 kHz, the value of C_R is:

$$C_R \approx \frac{1}{6.28 \times 200 \text{ kHz} \times 300} \approx 0.003 \ \mu\text{F}$$

With a value of 30K for R_N, 100K for R_F, and an F_3 of 400 kHz, the value of C_N is:

$$C_N \approx \frac{1}{6.28 \times 400 \text{ kHz} \times (30\text{K} + 100\text{K})} \approx 3 \text{ pF}$$

With a value of 0.003 μF for C_R, an R_F of 100K and an R_R of 300 ohms, the frequency of F_1 is:

$$F_1 \approx \frac{10}{6.28 \times 0.003 \ \mu\text{F} \times [100\text{K} + (10 \times 300)]} \approx 5 \text{ kHz}$$

With a value of 3 pF for C_N, an R_F of 100K and an R_N of 30K, the frequency of F_4 is:

$$F_4 \approx \frac{40}{6.28 \times 3 \text{ pF} \times [(40 \times 30\text{K}) + 100\text{K}]} \approx 1.6 \text{ MHz}$$

2-12. FEED-FORWARD TECHNIQUE

The capabilities of an op-amp can be increased by means of the feed-forward technique. The basic principle in the feed-forward concept is to extend the power bandwidth by routing the high frequencies around the first stages of the op-amp through a booster amplifier. Some op-amps have provisions for the feed-forward technique, or can be adapted to accommodate the idea. The only requirement is that the input to the final stage of the op-amp be accessible.

Figure 2-17 illustrates the feed-forward technique using a Motorola MC1539 op-amp. The input for the final stage of the MC1539 is accessible at terminal 5. An external booster amplifier is connected in parallel with the first two stages of the op-amp. The high frequency booster

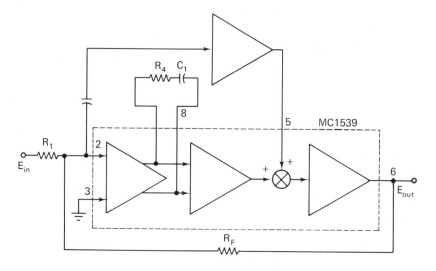

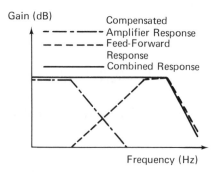

Fig. 2-17. Basic feed-forward circuit (Courtesy Motorola).

amplifier is designed to complement the natural rolloff of the compensated closed-loop op-amp. As shown by the graph, the feed-forward (booster) response increases as frequency increases. This compensates for the decrease in gain for frequency increases through the op-amp. The high frequency booster takes over completely when the input frequency is too high for the input stage to respond. The final rolloff (of E_{OUT}) is set by the characteristics of the op-amp output stage, which must pass all frequencies, and the booster amplifier rolloff characteristics.

Figure 2-18 illustrates the MC1539 op-amp with the feed-forward amplifier in place. (The characteristics of the feed-forward amplifier are illustrated in Fig. 2-19.) With the configuration shown in Fig. 2-18, a

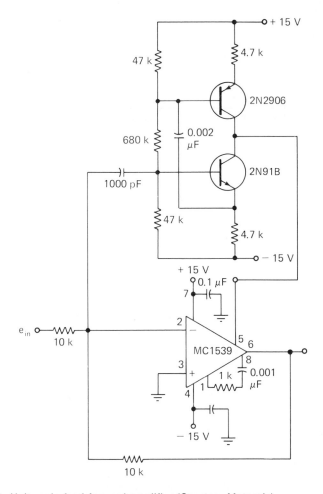

Fig. 2-18. Unity gain feed-forward amplifier (Courtesy Motorola).

10V (peak-to-peak) unity-gain output can be obtained at frequencies between 1 and 2 MHz. Without feed-forward, the MC1539 will produce a 10V (peak-to-peak) unity-gain output up to about 2 kHz. Thus, the unity-gain power bandwidth is extended by a decade.

The feed-forward amplifier shown in Fig. 2-19 can be used with any op-amp operating in the same general frequency range. The circuit will

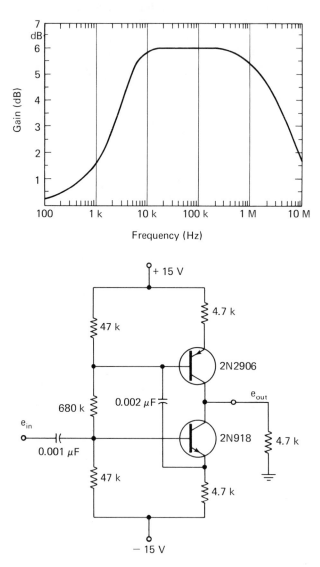

Fig. 2-19. Feed-forward amplifier response (Courtesy Motorola).

operate with power supplies as low as 6V, instead of the 15V shown. Of course, the peak-to-peak output is limited by the power supply voltages.

2-13. EXTERNAL POWER OUTPUT

The output power capability of an op-amp can be increased by the addition of an external emitter-follower stage, or a class B push-pull output stage. The emitter-follower approach is highly inefficient from a dissipation standpoint. The class B power stage, as shown in Fig. 2-20, is much more efficient.

In the circuit of Fig. 2-20, the diodes should have the same voltage drop as the base-emitter junction of the transistors (typically 0.5 to 0.7V for silicon). The values of the resistors in series with the diodes should be sufficient to drop the supply voltages to the nominal 0.5V level. Of

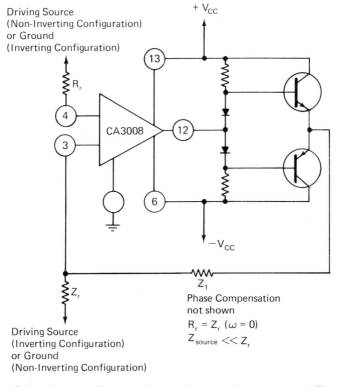

Fig. 2-20. External push-pull output stage to increase the output capability of an op-amp (Courtesy RCA).

course, the amount of current flowing through these resistors is dependent upon many factors such as transistor base current, forward current of the diodes and so forth. However, as a trial value, assume that the current flow is 10 mA or less.

The circuit of Fig. 2-20 is not suited for open-loop operation. The closed-loop resistances Z_r and Z_f should always be used. Under open-loop conditions where there is no feedback, the output transistors are subject to "thermal runaway." If an op-amp must be used in the open-loop condition and the output power must be increased, use a conventional power amplifier with emitter feedback. Such amplifiers are described in the author's *Handbook of Simplified Solid State Circuit Design* (Prentice-Hall, Inc., Englewood Cliffs, N. J., 1971).

2-14. EXTERNAL INPUT EMITTER FOLLOWERS

In some cases in which an op-amp must be driven from a high impedance source, or a source that cannot tolerate high levels of direct current, it may be necessary to reduce the input bias current. Of course, the input bias current of the op-amp can not be changed. However, the apparent input bias current is substantially decreased when emitter followers are added to the input terminals of the op-amp.

An emitter follower modification to an op-amp is illustrated in Fig. 2-21. With the particular op-amp shown, the apparent input bias current is

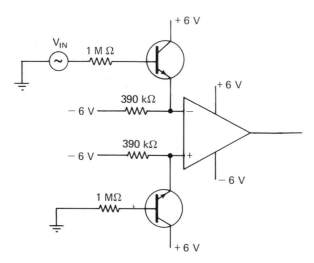

Fig. 2-21. Addition of input emitter followers to reduce input bias current requirements of an op-amp.

reduced from about 5000 pA (the nominal input bias of the IC op-amp) to about 140 pA per input. It should be noted that this modification also results in a decrease in bandwidth. For the op-amp shown in Fig. 2-21, the bandwidth is decreased almost a decade. That is, without modification, the op-amp output drops to zero at about 20 MHz. With addition of the emitter followers, the op-amp output is zero at about 2 MHz.

Keep in mind that there are many op-amps with a high input impedance, and low input bias current. It is generally preferable to use such op-amps, and not use external modification circuits. However, if an op-amp must be used for some other reason, the circuit of Fig. 2-21 will increase input impedance (to about 1 megohm) and reduce input bias current.

2-15. VOLTAGE-TO-CURRENT CONVERTER (TRANSADMITTANCE AMPLIFIER)

Figure 2-22 is the working schematic of an op-amp used as a voltage-to-current converter. The circuit is also known as a transadmittance amplifier, and sometimes as a current feedback amplifier. The circuit is used to supply a current (to a variable load) that is proportional to the voltage applied at the input of the op-amp (rather than proportional to the load). The current supplied to the load is relatively independent of the load characteristics.

Current sampling resistor R is used to provide the feedback to the positive (noninverting) input. When R_1, R_2, R_3 and R_4 are all the same value, the feedback maintains the voltage across R at the same value as the input. When R_2 is made larger than R_1, the voltage across R remains constant at a value equal to the ratio R_2/R_1.

If a constant input voltage is applied to the op-amp, the voltage across R also remains constant, regardless of the load (with very close tolerances). If the voltage across R remains constant, the current through R must also remain constant. With R_3 and R_4 normally much larger than the load impedance, the current through the load remains nearly constant, regardless of a change in impedance.

The most satisfactory configuration for the circuit of Fig. 2-22 is that in which the op-amp is operated as a unity gain amplifier, with the values of R_1 through R_4 all the same. This requires that the input voltage be sufficient to produce the desired current (or power) for the load. If the input voltage or signal is not sufficient, the values of R_1 and R_2 must be selected to provide the necessary gain.

The value of R must be selected to limit the output power [$I^2 \times (R + $ load)] to a value within the capability of the op-amp. For example, if the op-amp is rated at 700 mW total dissipation, with 100 mW dissipation for the basic op-amp, the total output power must be limited to 600 mW. As a guideline, make the value of R approximately one-tenth of the load.

2-15.1 Design example

Assume that the circuit in Fig. 2-22 is to be used as a voltage-to-current converter. The output load Z is 30 ohms (nominal), but is subject to variation. The maximum power output of the op-amp is 660 mW. It is desired to maintain the maximum output current, regardless of variation in output load, with a constant input voltage of 7 mV.

With a 30 ohm load, the value of R is approximately 3 ohms. The combined resistance of R and the load Z is then 33 ohms.

With a total resistance of 33 ohms, and a maximum power output of 660 mW for the op-amp, the maximum possible output current is:

$$\sqrt{0.660/33} = \sqrt{0.02} = 0.14 \text{ A (140 mA)}$$

With a value of 30 ohms for R, and 0.14A through R, the drop across R is 0.42 V (420 mV).

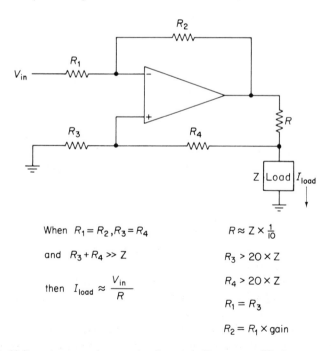

When $R_1 = R_2, R_3 = R_4$ $R \approx Z \times \frac{1}{10}$

and $R_3 + R_4 \gg Z$ $R_3 > 20 \times Z$

then $I_{load} \approx \dfrac{V_{in}}{R}$ $R_4 > 20 \times Z$

     $R_1 = R_3$

     $R_2 = R_1 \times \text{gain}$

Fig. 2-22. Voltage-to-current converter (transadmittance amplifier).

With 420 mV required at the output, and 7 mV at the input, the required amplifier gain is: 420/7 = 60.

The values of R_3 and R_4 must be at least 600 ohms (20 x 3 ohms) each, as shown by the equations. Any value above 600 ohms is satisfactory for R_3 and R_4.

The value of R_1 is the same as R_3, or at least 600 ohms. With a value of 600 ohms for R_1, and a gain of 60, the value of R_2 is: 600 ohms × 60 = 36,000 ohms. If values higher than 600 ohms are selected for R_1, to match higher values for R_3 and R_4, the value of R_2 must be increased accordingly.

2-16. VOLTAGE-TO-VOLTAGE CONVERTER

Figure 2-23 is the working schematic of an op-amp used as a voltage-to-voltage converter. The circuit is also known as a voltage gain amplifier. This circuit is similar to the voltage-to-current converter (Sec. 2-15) except that the load and the current sensing resistor are transposed. The voltage across the load is relatively independent of the load characteristics.

The most satisfactory configuration for the circuit of Fig. 2-23 is where the op-amp is operated as a unity gain amplifier, with values of R_1 through R_4 all the same. This requires that the input voltage be sufficient to produce the desired current (or power) for the load. If the input voltage or signal is not sufficient, the values of R_1 and R_2 must be selected to provide the necessary gain.

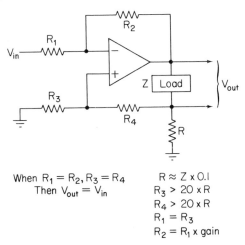

When $R_1 = R_2$, $R_3 = R_4$
Then $V_{out} = V_{in}$

$R \approx Z \times 0.1$
$R_3 > 20 \times R$
$R_4 > 20 \times R$
$R_1 = R_3$
$R_2 = R_1 \times gain$

Fig. 2-23. Voltage-to-voltage converter.

The value of R is selected to limit the output power $[I^2 \times (R + \text{load})]$ to a value within the capability of the op-amp. For example, if the op-amp is rated at 700 mW total dissipation, with 100 mW dissipation for the basic op-amp, the total output power must be limited to 600 mW. As a guideline, make the value of R approximately one-tenth of the load.

2-16.1 Design example

Assume that the circuit of Fig. 2-23 is to be used as a voltage-to-voltage converter. The output load is 90 ohms (nominal), but is subject to variation. The maximum power output of the op-amp is 600 mW. It is desired to maintain the maximum output voltage across the load, regardless of variation in load (within the current capabilities of the op-amp), with a constant input voltage. The available input signal voltage is any value up to 8V. The op-amp is capable of input and output voltages up to at least 8V without damage.

With a 90 ohm load, the value of R is approximately 9 ohms. The combined resistance of R and the nominal load is then 99 ohms.

With a total resistance of 99 ohms, and a maximum power output capability of 0.6W for the op-amp, the maximum possible output current is approximately 0.081A.

$$(\sqrt{0.6/99} \approx \sqrt{0.0066} \approx 0.081)$$

With a nominal value of 90 ohms for the load and 0.081A through the load, the maximum drop across the load is: $90 \times 0.081 = 7.29$ V. Since the op-amp is capable of an 8V input, and an 8V signal is available, adjust the input signal voltage to 7.29V.

The values of R_3 and R_4 should be at least 180 ohms (20×9 ohms) as shown by the equations of Fig. 2-23, with a value of 9 ohms for R.

The value of R_1 should be the same as R_3, or 180 ohms.

Since no gain is required, the value of R_2 should also be the same value as R_1, or 180 ohms. If values greater than 180 ohms are selected for R_3 and R_4, use the same values for R_1 and R_2. Thus, R_1 through R_4 will all be the same value.

2-17. HIGH-IMPEDANCE BRIDGE AMPLIFIER

Figure 2-24 is the working schematic of op-amps used as high-impedance bridge amplifiers. The circuit is not limited to use with a bridge. However, a bridge of any type will provide greater accuracy

when its output is fed to a high-impedance amplifier. Under these conditions, little or no current is drawn from the bridge, and the bridge
voltage output is amplified as necessary to provide a given reading.

Amplification for the entire circuit is provided by op-amp 3, and is
dependent upon the ratio of R_2/R_1. Op-amps 1 and 2 act as voltage
followers (Sec. 2-4), and provide no amplification. However, op-amps
1 and 2 do provide a high impedance for the bridge output (easily 100,000
ohms and above).

The value of R_1 is chosen on the basis of input bias current and voltage
drop, as described in Sec. 2-2.2 of this chapter. The value of R_2 is selected to provide the desired gain. Resistors R_3 and R_4 should be equal
to the values of R_1 and R_2, respectively.

2-17.1 Design example

Assume that the circuit of Fig. 2-24 is to provide about 8V
output. The bridge output is 500 mV. The op-amp 3 input bias is 200 nA.

The outputs of op-amps 1 and 2 are equal to (or slightly less) than the
bridge output, since op-amps 1 and 2 are connected as voltage followers
(Sec. 2-4). Thus, op-amp 3 must provide a gain of 16, for a circuit output
of 8V (8V/500 mV).

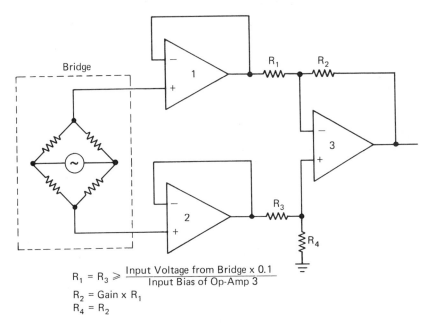

$$R_1 = R_3 \geqslant \frac{\text{Input Voltage from Bridge} \times 0.1}{\text{Input Bias of Op-Amp 3}}$$
$$R_2 = \text{Gain} \times R_1$$
$$R_4 = R_2$$

Fig. 2-24. High-impedance bridge amplifier.

The input voltage to op-amp 3 is 500 mV (or slightly less). The voltage drop across R_1, due to input bias current, should be no greater than 10 percent of the input voltage. With a drop of 50 mV across R_1, and a 200 nA input bias current, the value of R_1 is 250 kilohms, maximum. The value of R_3 should also be 250 kilohms maximum. Assume that values of 100K are used for R_1 and R_3.

With a value of 100K for R_1, and a required gain of 16, the value of R_2 is 1.6 megohms. R_4 should also be 1.6 megohms.

2-18. DIFFERENTIAL INPUT/DIFFERENTIAL OUTPUT

Figure 2-25 is the working schematic of two op-amps connected to provide differential-in/differential-out amplification. Generally, op-amps have a single-ended output, although some IC op-amps have a differential output.

There are cases in which a differential output is required. For example, the circuit of Fig. 2-25 can be used at the input of a system in which there is considerable noise pickup at the input loads, but the input differential signal is small. Since the noise signals are common mode (appear on both input leads simultaneously, at the same amplitude and polarity) they will not be amplified. However, the differential input is amplified, and appears as an amplified differential output voltage.

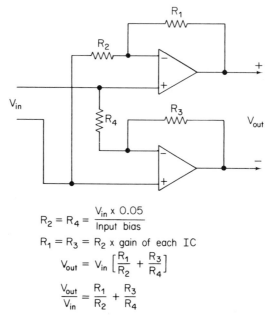

$$R_2 = R_4 = \frac{V_{in} \times 0.05}{\text{Input bias}}$$

$$R_1 = R_3 = R_2 \times \text{gain of each IC}$$

$$V_{out} = V_{in} \left[\frac{R_1}{R_2} + \frac{R_3}{R_4} \right]$$

$$\frac{V_{out}}{V_{in}} = \frac{R_1}{R_2} + \frac{R_3}{R_4}$$

Fig. 2-25. Differential input/differential output.

The circuit of Fig. 2-25 can be formed with two identical op-amps or, preferably, a dual-channel IC op-amp. The dual-channel IC is preferable since both channels will have identical characteristics (gain, bias, etc.). This is because both channels are fabricated on the same semiconductor chip. However, the circuit will work satisfactorily if the two op-amps are carefully matched as to characteristics (particularly in regards to input bias and input offset voltage).

As shown by the equations, each op-amp (or each channel) provides *one-half of the total differential output gain*. Thus, if each op-amp provides a gain of 10, the differential output is 20 times the differential input.

The output voltage swing is double that of the individual op-amps. Generally, the maximum output swing of a single-ended op-amp is slightly less than the $V_{CC} - V_{EE}$ voltage. Using the circuit of Fig. 2-25, the maximum differential output voltage swing is twice that of the $V_{CC} - V_{EE}$ voltage. For example, if the $V_{CC} - V_{EE}$ voltage is 15V, the maximum differential output is slightly less than 30V.

The values of R_2 and R_4 are chosen on the basis of input bias current and voltage drop, as described in Sec. 2-2.2. The values of R_1 and R_3 are chosen to provide the desired gain. Resistors R_3 and R_4 should be equal in value to R_1 and R_2, respectively.

2-18.1 Design example

Assume that the circuit of Fig. 2-25 is to provide a differential output with swing of approximately 30V. The available differential input is 30 mV. The input bias current is 200 nA.

With an input of 30 mV and a desired output of 30V, the circuit must provide a total gain of 1000 (30V/0.03V). Each op-amp must provide a gain of 500.

The voltage drop across R_2, due to input bias current, should be no greater than 5 percent of the input voltage. With a drop of 1.5 mV across R_2, and a 200 nA input bias current, the value of R_2 is: 1.5 mV/200 nA = 7500 ohms, maximum. R_4 should also be 7500 ohms.

With a value of 7500 ohms for R_2, and a required gain of 500, the value of R_1 is 3.75 megohms. R_3 should also be 3.75 megohms.

With a required differential output voltage swing of 30V, the $V_{CC} - V_{EE}$ voltage of the op-amps must be about 15V.

2-19. TEMPERATURE SENSOR

Figure 2-26 is the working schematic of an op-amp used as the major active element in a temperature sensor circuit. Temperature sensing with thermistors is popular because thermistors are inexpensive and easy to use. However, thermistors are nonlinear and can supply only very small output signals (generally a few microwatts).

The circuit of Fig. 2-26 overcomes these limitations. The output voltage is *relatively linear* over the temperature range of the op-amp. The output is *exactly linear* near the temperature at which the thermistor resistance equals the resistance of R_1. Thus, R_1 should be equal to R_T (the thermistor resistance) at the center of the desired temperature range.

The reference voltage V_{ref} is obtained from the Zener diode CR_1 and the voltage divider. The upper limit is a value determined by the power rating (P_T) of the thermistor, as shown by the equations. With the values shown, R_2 is adjusted so that V_{ref} is -0.067V. Under these conditions, the maximum power dissipated by the thermistor is about 2.5 μW, well under the 5 μW rating.

2-20. ANGLE GENERATOR

Figure 2-27 is the working schematic of two op-amps connected as an angle generator. As shown by the equations, the output of op-amp 1 is proportional to the sine of the input phase angle, and the output of op-amp 2 proportional to the cosine of the input phase angle.

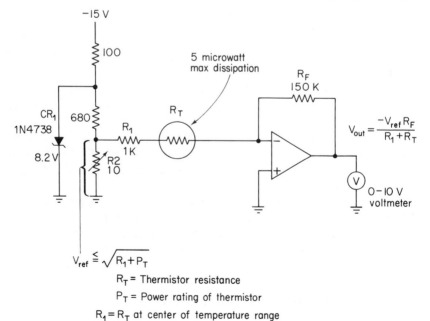

$$V_{ref} \leq \sqrt{R_1 + P_T}$$

R_T = Thermistor resistance

P_T = Power rating of thermistor

$R_1 = R_T$ at center of temperature range

Fig. 2-26. Temperature sensor.

The op-amps are connected as a Scott-T transformer into a three-wire synchro line.

When all resistor values are the same, the output of op-amp 2 is equal to twice the input voltage, multiplied by the cosine of the phase angle. For example, if the phase angle is 33°, and the input voltage (line-to-line three-phase input) is 1V, the op-amp 1 output is: $2 \times 1 = 2$; 2×0.8387 (the cosine of 33°) $= 1.6774$V.

With the same conditions (phase angle of 33°, input 1V) the op-amp 1 output is: 1.732×0.5446 (the sine of 33°) = approximately 0.94V.

The accuracy of the circuit in Fig. 2-27 is dependent upon matching of the op-amps, as well as matching of the resistors. A dual-channel IC op-amp is ideal for the circuit since both channels are fabricated on the same chip. However, two separate op-amps with closely-matched characteristics will produce satisfactory results. The resistors should have a tolerance of 1 percent (or better).

The circuit of Fig. 2-27 is most effective when the three-phase voltages are on the order of 1V, or a fraction of 1V (such as some analog computer

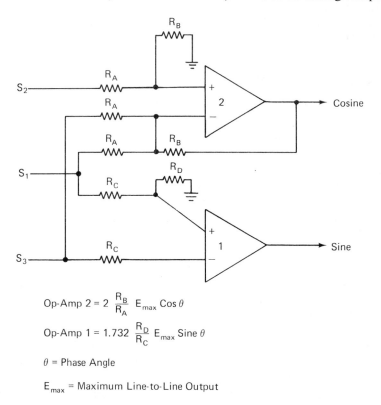

$$\text{Op-Amp 2} = 2 \ \frac{R_B}{R_A} \ E_{max} \ \text{Cos} \ \theta$$

$$\text{Op-Amp 1} = 1.732 \ \frac{R_D}{R_C} \ E_{max} \ \text{Sine} \ \theta$$

θ = Phase Angle

E_{max} = Maximum Line-to-Line Output

Fig. 2-27. Angle generator.

servo systems), and it is desired to have output readings in the 5V to 10V range (typical of digital logic systems).

The circuit in Fig. 2-27 has several advantages over direct measurement of the phase angle by meters. First, the op-amps present far less loading than a meter. (Of course, this is more important with low-voltage systems than with three-phase systems in the 120V range). Second, the circuit is independent of frequency. Most phase angle meters are for one frequency only. Third, the output voltage can be "weighted" or "scaled." For example, some design problems may require that the sine output be multiplied by 5, with the cosine multiplied by 10, or vice versa. This is accomplished by setting the gain of the individual op-amps to different levels (by different ratios of R_B/R_A and R_D/R_C).

The values of the input resistors R_A and R_D are chosen on the basis of input bias current and voltage drop, as described in Sec. 2-2.2. The *same value* should be used for all input resistors. The drop produced by input bias current across the input resistors should be 10 percent (or less) of the line-to-line input voltage. The values of R_B and R_D are chosen on the basis of desired gain for the individual outputs. The values of both R_B resistors should be the same, as should both R_D resistors. However, R_B need not equal R_D.

2-20.1 Design example

Assume that the circuit in Fig. 2-27 is to provide outputs from a three-phase system (S_1, S_2, S_3) with a maximum line-to-line voltage of 2V. The op-amp has an input bias current of 200 nA. The sine output should not exceed 3.5V. The cosine output should be 10V or less.

With a 2V input from the three-phase line, and a 200 nA input bias current, the value of each R_A and R_C resistor should be 1 megohm ($2V \times 0.1 = 0.2V$; $0.2V/200$ nA = 1 megohm).

Since the sine output is not to exceed 3.5V, op-amp 1 should have no gain. Thus, both R_D resistors should be 1 megohm. With no gain, the output from op-amp 1 will go from 0V, when the phase angle is 0°, to about 3.464V, when the phase angle is 90°.

Since the cosine output is not to exceed 10V, find the maximum output from op-amp with no gain (4V, or twice the input of 2V, in this case), then divide the maximum output limit by the no-gain output, or: 10/4 = 2.5. Thus, a gain of 2.5 is required for op-amp 2, and the values of both R_B resistors should be 2.5 megohm (1 M × 2.5 = 2.5 M).

To provide a 10V cosine output, V_{CC} and V_{EE} for op-amp 2 should be at least 10V (11 or 12V, in practical use). A lower value can be used for op-amp 2. However, a better match, and a more practical circuit, is obtained if both op-amps have the same $V_{CC} - V_{EE}$ voltage.

3. OPERATIONAL TRANSCONDUCTANCE AMPLIFIERS

An Operational Transconductance Amplifier (OTA) is similar in form to a conventional op-amp. However, OTAs and op-amps are not always interchangeable. For that reason an explanation of the unique characteristics found in OTAs is in order. The OTA not only includes the usual differential inputs of an op-amp but also contains an additional control input in the form of an *amplifier bias current* (or I_{ABC}). This control input increases the OTA's flexibility for use in a wide range of applications.

The characteristics of an ideal OTA are similar to those of an ideal op-amp except that the OTA has an extremely *high output impedance*. Because of this difference, the output signal of an OTA is best described in terms of *current that is proportional to the difference between the voltages of the two inputs*. Thus, the transfer characteristic is best defined in terms of *transconductance* rather than voltage gain. That is, transconductance $g_m = diff\ I_{out}/diff\ E_{in}$. Except for the high output impedance, and the definition of input/output relationships, the characteristics of a typical OTA are similar to those of a typical op-amp.

This chapter describes operation of the OTA and features various circuits using the OTA. As is discussed, the OTA provides the equipment designer with a wider variety of circuit arrangements than does the conventional op-amp. This is because the user can select the optimum circuit conditions for a specific application simply by varying the bias (I_{ABC}) conditions of the OTA. For example, if low power consumption, low bias and low offset current, or high input impedance are desired, then low I_{ABC} current may be selected. On the other hand, if operation into a moderate load impedance is the main consideration, then higher levels of I_{ABC} bias may be used.

117

3-1. CIRCUIT DESCRIPTION OF TYPICAL OTA

Figure 3-1 shows the equivalent circuit for the OTA. The output signal is a "current" that is proportional to the transconductance (g_m) of the OTA established by the amplifier bias current I_{ABC} and the differential input voltage. The OTA can either source or sink current at the output, depending upon the polarity of the input signal.

Figure 3-2 is a simplified block diagram of an OTA. Transistors Q_1 and Q_2 form the usual differential input amplifier found in most op-amps. The lettered circles (with arrows leading either into or out of the circles) indicate *current-mirrors*. The use of current-mirrors is essential to OTA operation. There are two basic types of current-mirror, as indicated in Figs. 3-3 and 3-4.

Figure 3-3 shows the basic type of current-mirror composed to two transistors, one of which (Q_2) is diode-connected. Because diode-connected transistor Q_2 is not in saturation, and is "active," the "diode" formed by the connection may be considered as a transistor with 100 percent feedback. Therefore, the base current still controls the collector current, as is the case in normal transistor action. That is, collector current I_C equals beta times base current I_b. If a current I_1 is forced into the diode-connected transistor Q_2, the base-to-emitter voltage will rise until

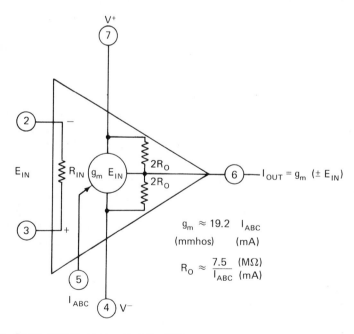

Fig. 3-1. Basic equivalent circuit of the OTA.

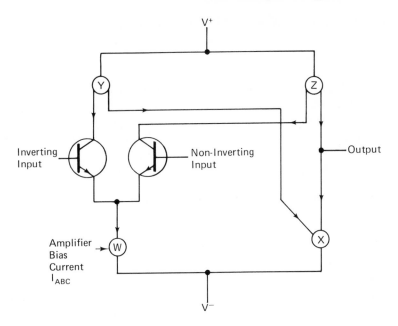

Fig. 3-2. Simplified diagram of OTA.

equilibrium is reached, and the total current being supplied is divided between the collector and base regions. Thus, a base-to-emitter voltage is established in Q_2 such that Q_2 "sinks" the applied current I_1.

If the base of a second transistor Q_1 is connected to the base-to-collector junction of Q_2, Q_1 will be able to "sink" a current approximately equal to that flowing in the collector lead of the diode-connected transistor Q_2. This assumes that both transistors have identical characteristics (which is usually the case with IC fabrication where all components are formed on the same chip).

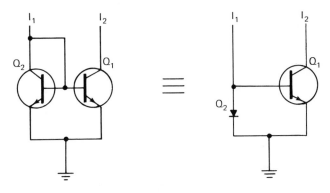

Fig. 3-3. Current mirror using diode connected transistor.

The difference in current between the input I_1 and the collector current I_2 of transistor Q_1 is due to the fact that the base-current for both transistors is supplied from I_1. The ratio of the "sinking" current I_2 to the input current I_1 is equal to I_2/I_1 = beta/(beta + 2). As beta increases, the output "sinking" current I_2 level approaches that of the input current I_1. Thus, the basic current-mirror of Fig. 3-3 is sensitive to transistor beta.

Circuit sensitivity to beta can be decreased, and an improvement in circuit output resistance characteristics can be made by the insertion of a diode-connected transistor in series with the emitter of Q_1. Such an arrangement is shown in Fig. 3-4. The diode-connected transistor Q_3 can be considered to be a current-sampling diode that senses the emitter current of Q_1 and adjusts the base current of Q_1 (via Q_2) to maintain a constant-current in I_2.

Current-mirror W in Fig. 3-2 uses the basic configuration of Fig. 3-3. Current-mirrors X, Y and Z are basically the constant-current version of Fig. 3-4. Mirrors Y and Z use PNP transistors, as shown by the arrows pointing outward from the mirrors.

Figure 3-5 is a complete schematic diagram of an OTA. The example shown is an RCA type CA3080 OTA available in IC form. The OTA shown uses only active devices (transistors and diodes, no resistors). Current applied to the amplifier-bias-current input I_{ABC} establishes the emitter current of the input differential amplifier Q_1 and Q_2. This provides effective control of the differential transconductance (g_m).

The g_m of a differential amplifier is equal to: $(q\ a\ I_c)/(2\ KT)$, where q is the charge on an electron, a is the ratio of collector current to emitter

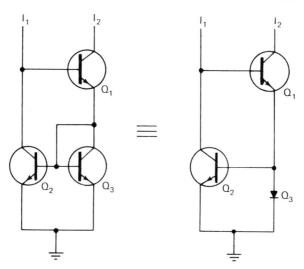

Fig. 3-4. Improved current mirror with additional transistor.

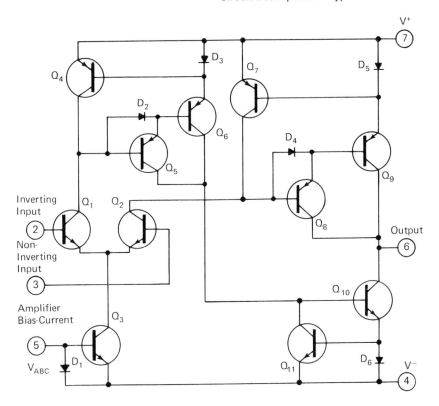

Fig. 3-5. RCA CA3080 and CA3080A operational transconductance amplifier (OTA) (Courtesy RCA).

current of the differential amplifier transistors (assumed to be 0.99 in this case), I_C is the collector current of the constant current source (I_{ABC} in this case), K is Boltzman's constant and T is the ambient temperature in degrees Kelvin.

In the case of the RCA CA3080, the transconductance $g_m = 19.2 \times I_{ABC}$, where g_m is in millimhos (mmho) and I_{ABC} is in mA. The temperature coefficient of g_m is approximately $-0.33\%/°C$ (at room temperature).

Transistor Q_3 and diode D_1 of Fig. 3-5 constitute the current-mirror W of Fig. 3-2. Similarly, transistors Q_7, Q_8 and Q_9, and diode D_5, make up the current-mirror Z of Fig. 3-2. Darlington-connected transistors are used in mirrors Y and Z to reduce the voltage sensitivity of the mirror, by an increase in mirror output impedance.

Transistors Q_{10}, Q_{11} and diode D_6 of Fig. 3-5 compose the current mirror X of Fig. 3-2. Diodes D_2 and D_4 are connected across the base-emitter junctions of Q_5 and Q_8, respectively, to improve the circuit speed. The amplifier output signal is derived from the collectors of the

Z and X current-mirror of Fig. 3-2, providing a push-pull Class A output stage that produces full differential g_m.

Figure 3-6 is the schematic diagram of another RCA OTA. This device, available in IC form, is designated as the CA3060. The OTAs in the CA3060 family incorporate a unique Zener diode regulator system (D_4, D_5, Q_{10}) that permits current regulation below supply voltages normally associated with such systems.

3-2. DEFINITION OF OTA TERMS

The following terms apply to all types of OTA circuits. However, the terms were first applied to IC OTA devices developed by RCA.

Amplifier Bias Current (I_{ABC})—The current supplied to the amplifier bias terminal to establish the operating point (such as the I_{ABC} current at the base of Q_3 in Fig. 3-6).

Amplifier Supply Current (I_A)—The current drawn by the amplifier from the positive supply source. The total supply current—which

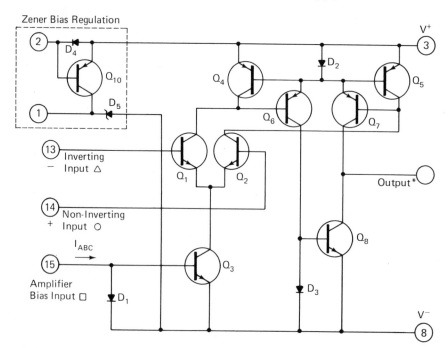

Fig. 3-6. Simplified diagram of OTA with bias regulator (Courtesy RCA).

includes the sum of the amplifier supply current, the amplifier bias currents and the regulator bias current—is not to be mistaken for the amplifier supply current.

Bias Regulator Current—The current flowing from the Zener bias regulator (such as at terminal 2 of Fig. 3-6), set by an external source, which establishes the operating conditions of the bias regulator.

Bias Terminal Voltage (V_{ABC})—The voltage existing between the amplifier bias terminal and the negative supply voltage terminal (such as between the I_{ABC} terminal and terminal 8 of Fig. 3-6).

Peak Output Current (I_{OM})—The maximum current that is drawn from a short circuit on the output of the amplifier (positive I_O) or the maximum current delivered into a short circuit load (negative I_O). Peak-to-peak current swing is twice the peak output current (I_{OM}).

Peak Output Voltage (V_{OM})—The maximum positive voltage swing $(V_{OM}+)$ or the maximum negative voltage swing $(V_{OM}-)$ for a specific supply voltage and amplifier bias.

Power Consumption (P)—The product of the sum of the supply voltages and the supply current, or $(V+ + V-) \times (I_A)$. This is not the total power consumed by an operating circuit. The power in the regulator must also be included for total power consumed.

Zener Regulator Voltage V_Z—The regulator voltage (such as across terminals 1 and 8 of Fig. 3-6), measured with current flowing in the bias regulator.

3-3. EFFECTS OF CONTROL BIAS ON OTA CHARACTERISTICS

Unlike conventional op-amps, the characteristics of OTAs can be altered by adjustment of amplifier bias input (I_{ABC}). In effect, many of the OTA characteristics can be programmed to meet specific design problems. The following is a summary of the effects of bias on typical OTA units. Although the absolute values given in the following examples apply to RCA OTAs, the range of characteristic changes shows the possible effects on operation of any OTA with variations in I_{ABC}. (Note that the characteristics listed here for OTAs are the same as for conventional op-amps described in Chapter 1.)

Input Offset Voltage V_{io} is not drastically affected by variations in I_{ABC}, as shown in Fig. 3-7. A possible exception is when the OTA is operated at high temperatures.

Input Offset Current I_{io} is directly affected by I_{ABC}, as shown in Fig. 3-8. For example, at $+25°C$, I_{io} increases almost in direct proportion with increases in I_{ABC}.

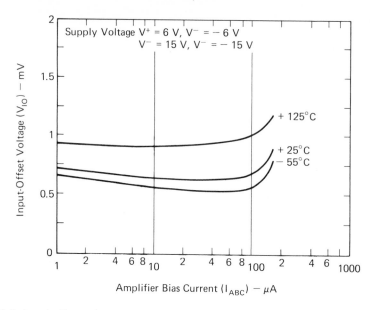

Fig. 3-7. Input offset voltage versus amplifier bias current for OTA (Courtesy RCA).

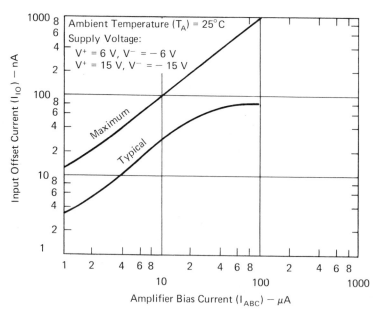

Fig. 3-8. Input offset current versus amplifier bias current for OTA (Courtesy RCA).

Input Bias Current I_{ib} also increases directly with increases in I_{ABC}, as shown in Fig. 3-9.

Peak Output Current I^+_{OM} or I^-_{OM} is another characteristic that increases with I_{ABC}, as shown in Fig. 3-10.

Peak Output Voltage V_{OM} is not drastically affected by variations in I_{ABC}, as shown in Fig. 3-11. Instead, V_{OM} is set (primarily) by supply voltage, as is the case with a conventional op-amp.

Amplifier Supply Current I_A, *Device Dissipation* P_D, and *transconductance* g_m all increase directly with I_{ABC}, as shown in Figs. 3-12, 3-13 and 3-14, respectively.

Input Resistance R_I and *Output Resistance* R_O both decrease with increases with I_{ABC}, as shown in Figs. 3-15 and 3-16, respectively.

Input and *Output Capacitance* C_I and C_O, as well as *Amplifier Bias Voltage* V_{ABC} all increase with I_{ABC}, as shown in Figs. 3-17 and 3-18, respectively. However, these characteristics do not increase in direct proportion to I_{ABC}. That is, a large increase in I_{ABC} occurs for a small increase in C_I, C_O and V_{ABC}.

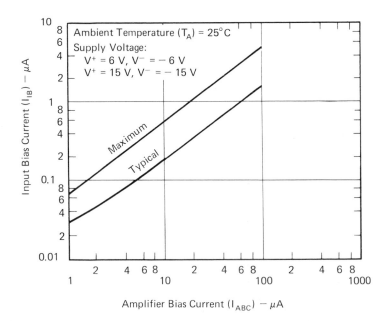

Fig. 3-9. Input bias current versus amplifier bias current for OTA (Courtesy RCA).

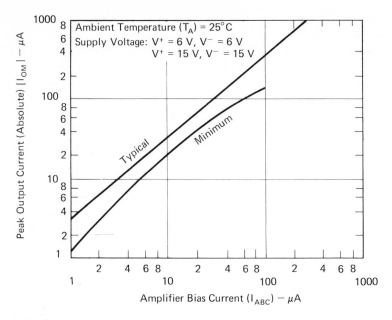

Fig. 3-10. Peak output current versus amplifier bias current for OTA (Courtesy RCA).

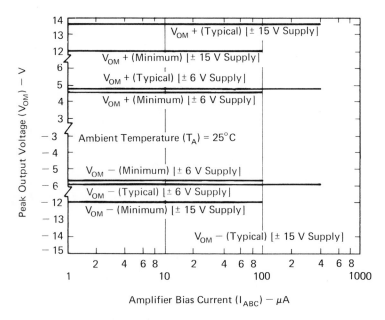

Fig. 3-11. Peak output voltage versus amplifier bias current for OTA (Courtesy RCA).

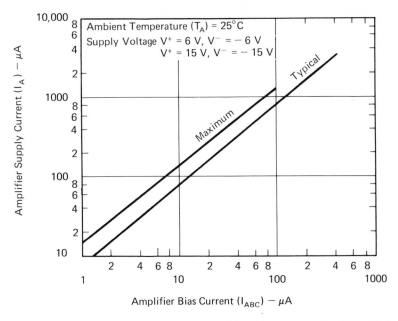

Fig. 3-12. Amplifier supply current versus amplifier bias current for OTA (Courtesy RCA).

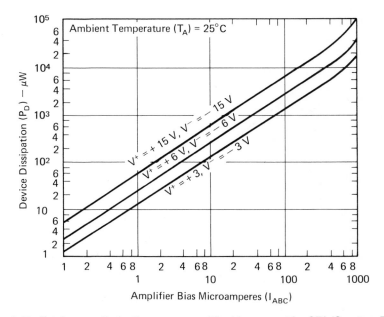

Fig. 3-13. Total power dissipation versus amplifier bias current for OTA (Courtesy RCA).

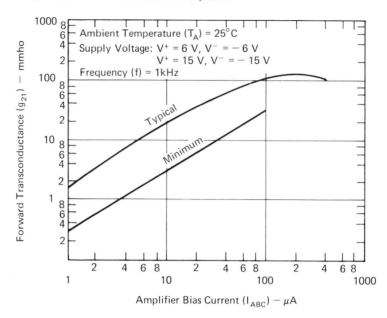

Fig. 3-14. Forward transconductance versus amplifier bias current for OTA (Courtesy RCA).

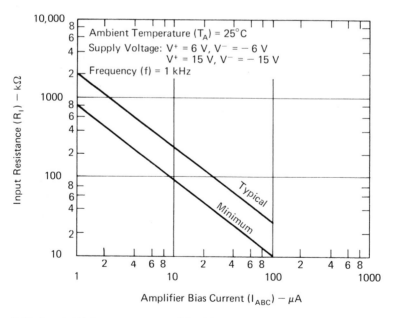

Fig. 3-15. Input resistance versus amplifier bias current for OTA (Courtesy RCA).

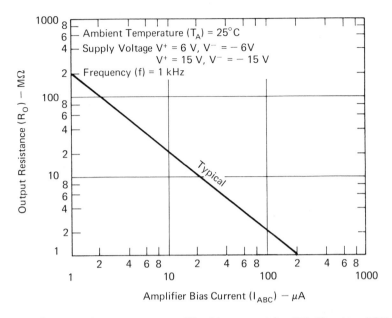

Fig. 3-16. Output resistance versus amplifier bias current for OTA (Courtesy RCA).

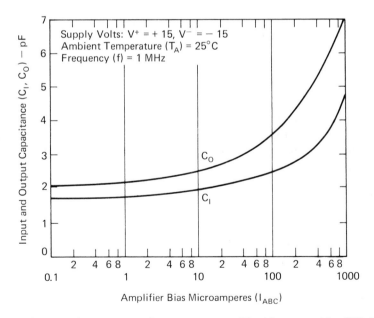

Fig. 3-17. Input and output capacitance versus amplifier bias current for OTA (Courtesy RCA).

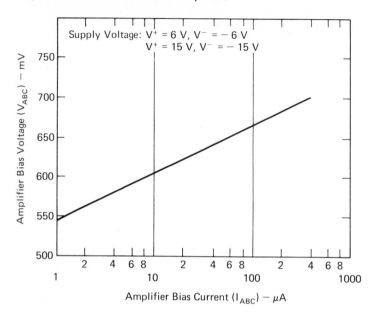

Fig. 3-18. Amplifier bias voltage versus amplifier bias current for OTA (Courtesy RCA).

3-4. BASIC DESIGN CONSIDERATIONS FOR OTA UNITS

The basic function of an OTA is as a substitute for an op-amp. However, the OTA allows the designer to select and control operating conditions of the op-amp circuit by adjusting the I_{ABC}. This permits the designer to have complete control over circuit transconductance, peak output current and total power consumption, relatively independent of supply voltage. In addition, the high output impedance makes the OTA ideal for applications in which current summing is involved.

The following steps outline procedures for design of an op-amp, using an OTA. The device selected for this example is a CA3060, which is an RCA OTA in IC form.

Figure 3-19 is the working schematic of a 20 dB op-amp using the CA3060. The circuit requirements are as follows:

Closed-loop voltage gain: 10 (20 dB)
Offset voltage: adjustable to zero
Current drain: as low as possible
Supply voltage: ±6V

Maximum input voltage: ±50 mV

Input resistance: 20K

Load resistance: 20K

As discussed in Chapter 1, the closed-loop gain is set by the ratio of feedback resistance R_F to input resistance R_S. With R_S specified as 20K, and a desired closed loop gain of 10, the value of R_F is: $10 \times 20K$ or 200K.

The next step is to calculate the required transconductance g_m (also listed as g_{21} on some OTA datasheets) to produce a suitable open-loop gain. Assume that the open-loop gain A_{OL} must be at least 10 times the closed-loop gain. With a closed-loop gain of 10, the open-loop gain must be: 10×10 or 100.

Open-loop gain A_{OL} is related directly to load resistance R_L and transconductance g_m. The required transconductance is equal to A_{OL} divided by R_L. With an A_{OL} of 100 and an R_L of 20K, the g_m should be: $100/20K =$ 5 millimhos (mmho). However, the actual load resistance is the parallel combination of R_L and R_F, or approximately 18K, found by $(20 \times 200)/(20 + 200) \approx 18$. With an A_{OL} of 100 and an actual load R_L of 18K, the g_m should be: $100/18K$, or approximately 5.5 mmho.

The transconductance g_m is set by I_{ABC}. With a datasheet curve similar to that of Fig. 3-14, select an I_{ABC} from the *minimum value* curve to

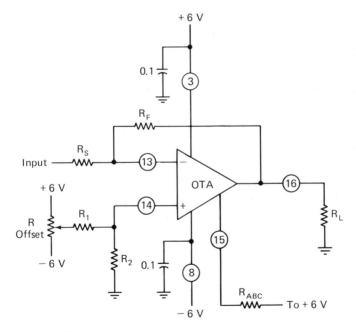

Fig. 3-19. Basic OTA system connections.

assure that the OTA will provide sufficient gain. As shown in Fig. 3-14, for a g_m (shown as g_{21}) of 5.5 mmho, the required I_{ABC} is approximately 20 μA.

Before calculating the value of R_{ABC} that will produce the desired I_{ABC}, check that the calculated I_{ABC} (20 μA) will produce the desired output swing capability. With an input of $\pm$50 mV, and a gain of 10, the output voltage swing is $\pm$0.5V, and 0.5V will appear across the output load. As discussed, the output load is equal to the parallel combination of R_L and R_F, or approximately 18K. With a 0.5V swing, and an approximate load of 18K, the total amplifier current output is approximately: 0.5V/18K, or 27.7 μA.

With a datasheet curve similar to that of Fig. 3-10, use the minimum value curve to check that an I_{ABC} of 20 μA will produce an I_{OM} of at least 27.7 μA. As shown in Fig. 3-10, for an I_{ABC} of 20 μA, the I_{OM} is approximately 40 μA, well above the desired minimum of 27.7 μA.

Once assured that the calculated I_{ABC} will meet the output swing requirements, calculate the value of R_{ABC}. As shown in Fig. 3-19, R_{ABC} is connected to the $+6$V supply. (R_{ABC} can be connected to the Zener bias regulator. However, this will increase current drain on the supply.)

As shown in Fig. 3-6, with R_{ABC} connected to $+V$, R_{ABC} and diode D_1 are in series between the $+V$ and $-V$ supplies, and there is a total of 12V across the series components. The drop across D_1, which is V_{ABC}, can be found by reference to a curve similar to Fig. 3-18. As shown in Fig. 3-18, with an I_{ABC} of 20 μA, V_{ABC} is approximately 630 mV, or 0.63V. The drop across R_{ABC} is $12 - 0.63$V, or 11.37V. For a drop of 11.37V and an I_{ABC} of 20 μA, the value of R_{ABC} is: 11.37/20 μA = 0.568M, or 568K. Use the next lowest standard resistor of 560K, to assure that a minimum I_{ABC} of 20 μA will flow.

The final step is to calculate values for the input offset adjustment circuit R_{offset}, R_1 and R_2. To reduce the loading effect of the offset adjustment circuit on the power supply, the values should be selected in a manner similar to that of a conventional op-amp. For example, the value of R_2 should be approximately equal to the parallel combination of R_S and R_F. This will equalize currents between the inverting and noninverting inputs. Thus, R_2 should equal: $(20 \times 200)/(20 + 200)$, or approximately 18K.

With a datasheet curve similar to that of Fig. 3-8, find the input offset current. As shown in Fig. 3-8, for an I_{ABC} of 20 μA, the input offset current should be a maximum of 200 nA. With 200 nA flowing through R_2, the voltage across R_2 is: $200^{-9} \times 18^{+3}$, or 3.6 mV. This 3.6 mV must be added to the maximum input offset voltage possible for the OTA. The OTA datasheet shows a maximum input offset voltage of 5 mV. Thus, the maximum voltage required at the noninverting input is: 5 mV + 3.6 mV or 8.6 mV.

The current necessary to provide a possible offset voltage of 8.6 mV across an 18K resistance is: $8.6^{-3}/18^{+3}$, or 0.48^{-6} (0.48 μA). This current must flow through R_1.

A possible $\pm$6V is available from R_{offset} to R_1. However, for a more stable circuit, assume that $\pm$1V is available to R_1. With 1V available, and a required current of 0.48 μA, the value of R_1 is: $1/0.48^{-6}$, or approximately 2M. Use the next larger standard value of 2.2M.

The value of R_{offset} is not critical. A larger value of R_{offset} will draw less current from the supplies. As a guideline, the maximum value of R_{offset} should be less than twice the value of R_1, or less than 4.4M in this case. Use a standard value of 4M.

This completes the final step in design. Other design considerations for OTAs are essentially the same as for conventional op-amps, as described in Chapter 1. For example, phase compensation is accomplished by the usual open-loop or closed-loop methods. (The datasheet for the CA3060 recommends open-loop compensation by means of a resistor and capacitor across the differential input terminals.) As always, the user should follow all datasheet recommendations. If none are available, then use one of the phase compensation methods described in Chapter 1. This same recommendation applies to selection of decoupling capacitor values.

3-4.1 Effects of capacitance on OTA circuit design

One major difference between the OTA and conventional op-amps is the high output impedance of all OTA units. A typical OTA will have a much higher output impedance than a comparable op-amp, even though the output impedance can be varied by adjustment of I_{ABC}. Because of this higher impedance, the effects of capacitance are much greater. For example, a 10K output load with a stray capacitance of 15 pF has an RC time constant of about 1 MHz. In designing any OTA circuit, particularly where the OTA is used as a feedback amplifier, stray capacitance must always be considered because of the adverse effect on frequency response and stability.

Figure 3-20 illustrates how a 10K, 15 pF load modifies the frequency characteristics of the RCA CA3060 unit.

Capacitive loading also has an effect on slew rate. The maximum slew rate is limited to the maximum rate at which the capacitance can be charged by the peak output current I_{OM}. (I_{OM} is set by I_{ABC} as shown in Fig. 3-10.)

As discussed in Chapter 1, slew rate is the difference in output voltage divided by the difference in time, and is expressed as dV/dt. In the case of an OTA connected as shown in Fig. 3-19, the slew rate is equal to the

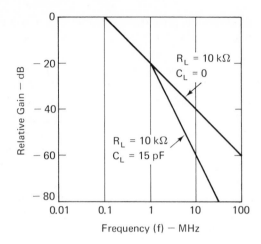

Fig. 3-20. Effect of capacitive loading on frequency response (Courtesy RCA).

I_{OM} divided by C_L, where C_L is the total load capacitance, including strays.

Figure 3-21 illustrates the relationship between slew rate and total load capacitance. For example, with a given peak output current I_{OM} of 10 μA and a load capacitance of 100 pF, the slew rate is 0.1V/μS. If the load capacitance can be decreased to 1 pF, the slew rate is increased 10V/μS.

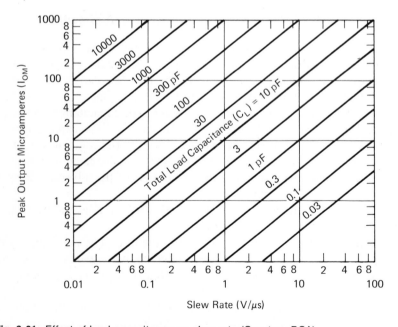

Fig. 3-21. Effect of load capacitance on slew rate (Courtesy RCA).

3-5. TYPICAL OTA APPLICATIONS

The following paragraphs describe how OTA units can be used. Keep in mind that an OTA can be used as a substitute for a conventional op-amp in any application, provided the OTA has comparable characteristics (frequency response, power output, etc.). The area in which an OTA cannot be substituted for a conventional op-amp is one in which low output impedance is required. Even when OTA output impedance is reduced to the minimum by adjustment of I_{ABC}, the impedance is generally higher than for a comparable op-amp.

3-5.1 OTA multiplexer

Because an OTA has the variable bias (I_{ABC}) feature, the OTA can be gated for multiplex applications. That is, the OTA can be gated full-on or full-off by means of pulses applied to the I_{ABC} input. In a multiplex circuit, two or more OTA units are connected so that their outputs are summed together, with the inputs receiving signals from separate sources. The resultant output is the combination of the input signals.

Figure 3-22 shows a simple two-channel multiplex system using two RCA CA3080 OTA devices. In this example, positive and negative 5V power supplies are used for the OTAs. The IC flip-flop is powered by the positive supply. If necessary to satisfy the logic supply requirements, the negative supply voltage may be increased to −15V, with the positive supply at +5V.

Outputs from the clocked flip-flop are applied through PNP transistors to gate the OTAs (on alternate half-cycles) at the I_{ABC} input. The transistors are connected in the grounded-base configuration to minimize capacitive feed-through (from flip-flop to OTA) via the base-collector junction of the PNP transistors.

There is some level shift between input and output of the multiplex system. In the case of the OTA units shown the level shift is about 2 mV for the CA3080A and 5 mV for the CA3080. Of course, the level shift will depend upon characteristics of the OTA used. In general, the level shift is due to input offset voltage of the OTA, rather than gain. Typically, open-loop gain of the system is about 100 dB, with normal loading at the output. To further increase gain, and reduce the effects of load, it is possible to add a buffer and/or amplifier stage at the OTA multiplexed output. Such a system is described in later paragraphs of this chapter.

The circuit in Fig. 3-22 may require some phase compensation at the output. This can be done by means of a simple RC network as shown. The values for the RC phase-compensation network are dependent upon

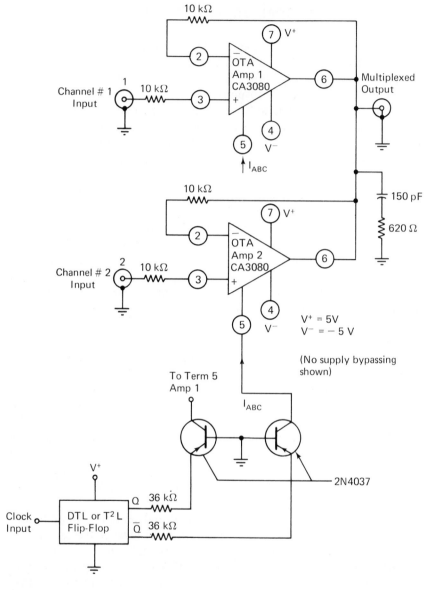

Fig. 3-22. Two-channel linear time-shared multiplex circuit using OTAs (Courtesy RCA).

frequency characteristics of the OTA. As a general rule, let 1/(6.28 RC) be approximately equal to the lowest frequency pole of the OTA. That is, let the *RC* network equal the lowest frequency at which OTA gain begins to drop. In the example of Fig. 3-22, the frequency is approximately 2 MHz. Refer to Chapter 1 for a further discussion of phase compensation.

Another two-channel multiplex system is shown in Fig. 3-23. This is similar to the multiplex system of Fig. 3-22, except the level-shifting transistors are omitted. These transistors are not needed in the circuit of Fig. 3-23 because of the higher output voltage from the flip-flop. The flip-flop of Fig. 3-23 is a COS/MOS unit with outputs in the order of ±10V.

Figure 3-24 shows a three-channel multiplex system using an RCA CA3060 and a 3N138. The CA3060 is an IC package that contains three identical OTA units. The 3N138 is a MOSFET used as a buffer/amplifier.

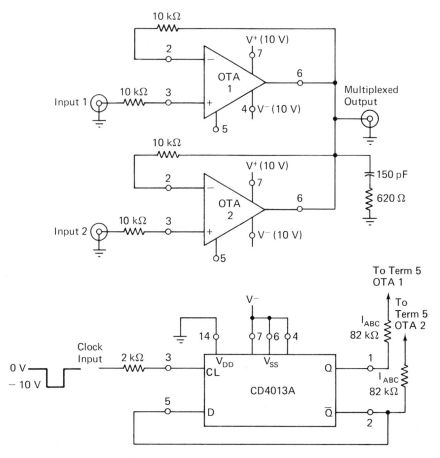

Fig. 3-23. Two channel linear multiplex system using COS/MOS flip-flop to gate two OTAs (Courtesy RCA).

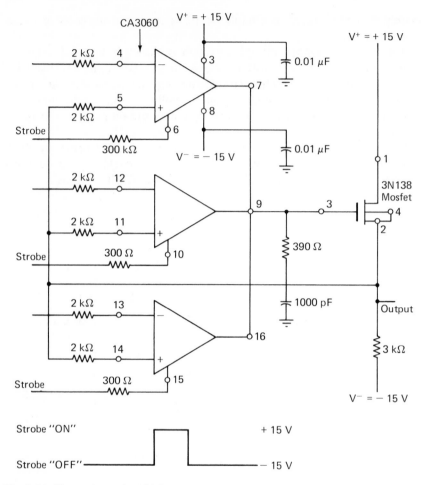

Fig. 3-24. Three-channel multiplexer.

Each OTA is connected as a high-input impedance voltage follower (similar to the op-amp voltage followers described in Chapter 2). A gating or "strobe" pulse is applied to each of the OTA units at some specific time interval. When the "strobe" pulse is applied, each OTA is activated, and the OTA output swings to the approximate level of the input (at that OTA unit). All three OTAs can be "strobed" simultaneously, if required. The resultant output is then the sum of the three OTA inputs.

The 3N128 MOSFET provides approximately 100 dB voltage gain at the output. Part of the output is fed back to each OTA unit. Without this feedback, gain will be unstabilized, and the system would not be

accurate. Generally, multiplex systems are designed so that the output is identical to the sum of inputs, or is at a level equal to the sum of inputs multiplied by a fixed amount of gain (such as sum of input times 100 dB).
 The values shown in Fig. 3-24 are for operation with supply voltages of ±15V. The system can be operated at ±6V, with modifications. The 300K resistors connected between the "strobe" pulses and the I_{ABC} inputs must be reduced to about 100K. Likewise, the 3K drain resistance of the 3N128 must be reduced to about 1K. When reduced (±6V) supply voltages are used, it is possible to use other MOSFET amplifiers, such as the RCA 40811.

The phase compensation network consists of a single 390 ohm resistor and a 1000 pF capacitor, connected at the interface between the OTA outputs and the MOSFET input. With these values, the bandwidth of the system is about 1.5 MHz, and the slew rate is 0.3V/μS.

3-5.2 OTA comparator

Any OTA can be used as the amplifier in a comparator circuit. The RCA CA3060 (described in Sec. 3-5.1 for the multiplexer) is ideal for use in a tri-level comparator circuit since the CA3060 has three idential OTA units in one IC package.

Figure 3-25 shows the functional block diagram of a tri-level comparator. The circuit has three adjustable limits. If either the upper or lower limit is exceeded, the appropriate output is activated until the input signal returns to a selected intermediate limit. Tri-level comparators are particularly suited to many industrial control applications.

As shown in Fig. 3-25, two of the three amplifiers are used to compare the input signal with the upper-limit and lower-limit reference voltages. The third amplifier is used to compare the input signal with a selected value of intermediate-limit reference voltage. By appropriate selection of resistance ratios the intermediate-limit may be set to any voltage between the upper-limit and lower-limit values.

The output of the upper-limit and lower-limit comparator sets the corresponding upper- or lower-limit flip-flop. The activated flip-flop retains its state until the third comparator (intermediate-limit) initiates a reset function, thereby indicating that the signal voltage has returned to the intermediate-limit selected.

The full circuit diagram of the tri-level comparator is shown in Fig. 3-26. The flip-flops are made up of RCA CA3086 transistor arrays (in IC form). Discrete transistors can be used instead, with similar resistance values. Power is provided for the CA3060 by ±6V supplies. The built-in regulator (see Fig. 3-6) provides amplifier bias current (I_{ABC}) to the three amplifiers.

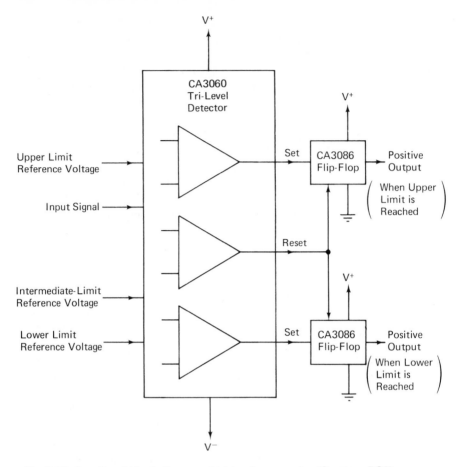

Fig. 3-25. Functional block diagram of tri-level comparator (Courtesy RCA).

In the circuit of Fig. 3-26, upper-limit and lower-limit reference voltages are selected by appropriate adjustment of potentiometers R_1 and R_2, respectively. When resistors R_3 and R_4 are equal in value (as shown), the intermediate-limit reference voltage is automatically established at a value midway between the lower-limit and upper-limit values. Other values of intermediate-limit voltages can be selected by adjustment of R_3 and R_4. The input signal E_S is applied to the three comparators through corresponding 5.1K resistors. The comparator outputs on the upper-limit and lower-limit SET lines trigger the appropriate flip-flop whenever the input signal reaches a limit value. When the input signal returns to an intermediate value, the common flip-flop RESET line is energized.

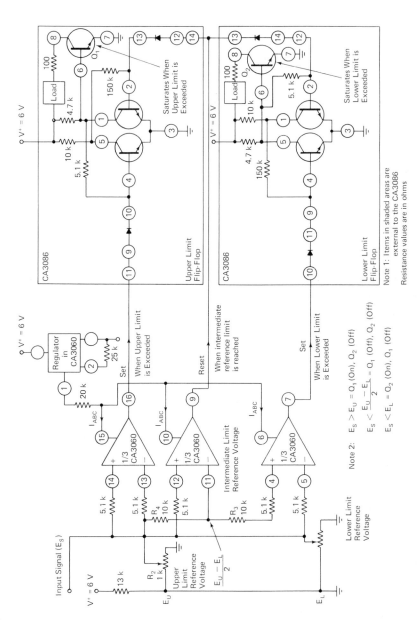

Fig. 3-26. Tri-level comparator circuit using OTAs (Courtesy RCA).

Note 1: Items in shaded areas are external to the CA3086. Resistance values are in ohms

Note 2: $E_S > E_U = Q_1$ (On), Q_2 (Off)

$E_S < \dfrac{E_U - E_L}{2} = Q_1$ (Off), Q_2 (Off)

$E_S < E_L = Q_2$ (On), Q_1 (Off)

The loads shown in the circuit of Fig. 3-26 are 5V, 25mA lamps. Other loads can be used, within the current and voltage capabilities of the flip-flop components.

3-5.3 OTA sample-and-hold circuits

The multiplex system described in Sec. 3-5.1 can be modified to produce sample-and-hold functions, using the "strobe" or gating characteristics of an OTA. That is, the I_{ABC} input of the OTA is pulsed to switch the OTA on and off. The output of the OTA then represents a "sample" of the OTA signal input (taken during the "on" period). The "hold" function is provided by means of a capacitor at the OTA output.

Figure 3-27 shows the basic sample-and-hold system using an RCA CA3080A as the amplifier. In this circuit, the OTA functions as a simple voltage-follower, with the phase-compensation capacitor C serving the additional function of sampled-signal storage. When the I_{ABC} input is at 0V (SAMPLE), the OTA is on. Under these conditions, the OTA output is at the same level as the signal input, and the 300 pF phase-compensation capacitor C charges to the level of the OTA output. When the I_{ABC} input is at −15V (HOLD), the OTA is off. However, the capacitor C remains charged.

The main problem with any sample-and-hold system using a charging capacitor is that the capacitor may discharge through leakage during the off or HOLD cycle. Such leakage could occur through the amplifier output circuit, or the 3N138 input circuit. However, since an OTA has very high output resistance, the leakage path through the OTA is practically nil. (The CA3080 has an output resistance in excess of 1000M under cut-off conditions.) Likewise, the gate leakage of the 3N138 is very low (typically 10 pA) since the transistor is a MOSFET (with insulated gate).

The open-loop voltage gain of the system is approximately 100 dB. The open-loop output impedance of the 3N138 is approximately 220 ohms (with a g_m of about 4600 μmho at an operating current of 5 mA). The system closed-loop output impedance is approximately equal to Z_O (open-loop output impedance of 3N138) divided by open-loop voltage gain, or 220 ohms/100 dB = 220 ohms/10^5 = approximately 0.0022 ohm. This output impedance is comparable to the closed-loop output impedance of a conventional op-amp.

Figure 3-28 shows a "sampled" triangular wave using the circuit of Fig. 3-27. The lower trace is the sampling signal (pulse applied to the

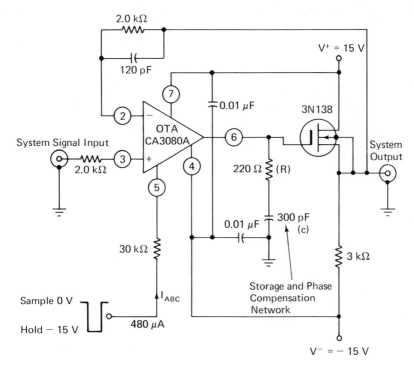

Fig. 3-27. Sample-and-hold circuit using an OTA (Courtesy RCA).

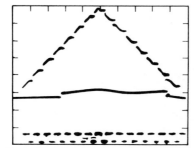

Top Trace: Sampled Signal 1 V/Div and 20 μsec/Div
Center Trace: Top Portion of Upper Signal
1 V/Div and 2 μsec/Div
Bottom Trace: Sampling Signal 20 V/Div and
20 μsec/Div

Fig. 3-28. Waveforms for sample-and-hold circuit (Courtesy RCA).

I_{ABC} input). When the sampling signal goes negative (to -15V), the OTA is cut off, and the system signal is "held" on the storage capacitor C, as shown by the plateaus on the triangular waveform. The center trace of Fig. 3-28 is a time expansion of the topmost transition (in the upper trace) with a time scale of 2 μS/div.

As discussed, variation in the stored signal level during the hold period is of concern in any sample-and-hold system using the charging capacitor method. This variation is primarily a function of the cutoff leakage current in the OTA output (a maximum limit of 5 nA), the leakage of the storage element (capacitor C), and other extraneous paths (such as gate leakage of the 3N138). The leakage currents may be either positive or negative. Consequently, the stored signal may rise or fall during the "hold" interval. The term "tilt" is used to describe this condition. Figure 3-29 shows the expected pulse "tilt" in microvolts as a function of time for various values of the compensation/storage capacitor C. The horizontal axis shows three scales representing typical leakage current of 50 nA, 5 nA and 500 pA.

As an example of how Fig. 3-29 can be used, assume that the hold period is 20 μS, and that capacitor C is 100 pF. If leakage can be limited to 500 pA, the pulse tilt will be 100 μV. That is, the system output level

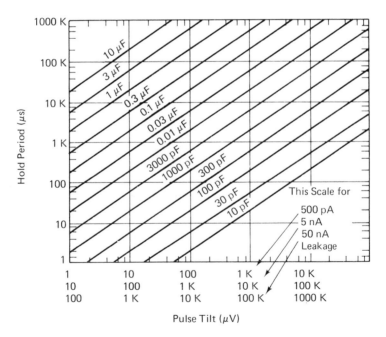

Fig. 3-29. Chart showing "tilt" in sample-and-hold potentials as a function of hold time with load capacitance as a parameter (Courtesy RCA).

could be shifted by 100 μV during the hold period. If the leakage current increases to 5 nA, the pulse tilt will increase to 1000 μV. The effects of level shift (or pulse tilt) are cumulative. For example, if 10 hold periods are required for a system signal input, the total level shift will be 10 times that for each hold period.

The circuit of Fig. 3-27 can be modified to operate with pulse levels found in DTL/TTL logic systems. Typically, DTL/TTL logic pulses are at 0V and 5V levels. Figure 3-30 shows the required modification. The control pulse to the I_{ABC} input is applied through a PNP transistor (RCA 2N4037) connected in the grounded-base configuration. Grounded-base is used to minimize capacitive feedthrough from the control pulse source to the I_{ABC} input. The 9.1K resistor connected to the 2N4037 emitter establishes I_{ABC} conditions similar to those used in the circuit of Fig. 3-27.

There is a tradeoff on the size of capacitor C. If the value of capacitor C is increased, larger signals can be sampled and held. However, each increase in capacitor C decreases slew rate. In general, use the largest possible capacitor C compatible with required slew rate. Figure 3-31

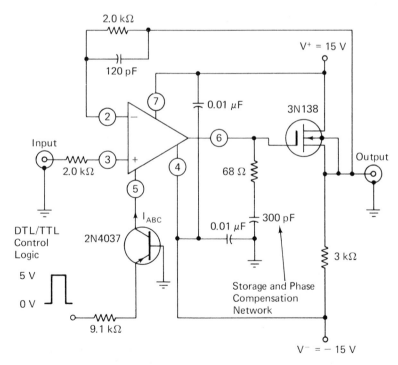

Fig. 3-30. Sample-and-hold circuit for DTL/TTL control logic using OTA (Courtesy RCA).

shows slew-rate as a function of I_{ABC} and various capacitors C. The information of Fig. 3-31 applies to capacitor C in both Figs. 3-27 and 3-30. The magnitude of the current being supplied to capacitor C is equal to the I_{ABC} when the OTA is supplying its maximum output current. As in the case with the multiplexer (Sec. 3-5.1) let the RC network equal the lowest frequency at which OTA gain begins to drop.

3-5.4 OTA gyrator

One of the problems in designing filters used at very low frequencies is the need for a very large inductance. This problem can be overcome by means of an active filter gyrator circuit. A gyrator is a circuit that appears as a variable inductance of high value (typically in the kilohenry region). However, the gyrator does not contain any inductive components.

Figure 3-32 shows a gyrator circuit composed of two OTA units. The circuit appears as an inductance of up to 10 kilohenry across terminals A and B. The circuit can be tuned to exact inductance values by means of the 100K tuning potentiometer R_1. The setting of R_1 varies the I_{ABC} current of both OTAs simultaneously. When one I_{ABC} current is increased, the opposite I_{ABC} current is decreased. This action varies the

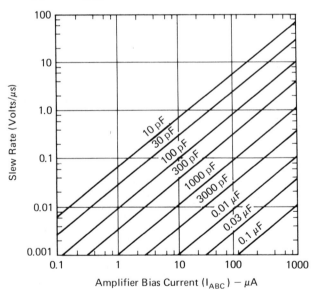

Fig. 3-31. Slew rate as a function of amplifier bias current, with phase-compensation capacitance as a parameter, for OTA (Courtesy RCA).

transconductance of both OTAs simultaneously, and serves to "tune" the circuit by changing the resistance of the OTAs.

The circuit of Fig. 3-32 makes use of the high impedance output available in OTA units. The Q of the 10 kilohenry variable inductance is approximately 13, using the circuit values shown. The 20K to 2M attenuators in this circuit extend the dynamic range of each OTA by a factor of 100.

3-5.5 OTA gain control and modulation circuits

An obvious function of an OTA is that of a gain control element. The gain of a signal passing through an OTA can be controlled by variation of the I_{ABC}. This is because transconductance of the OTA (and thus gain of the circuit) is directly proportional to I_{ABC}.

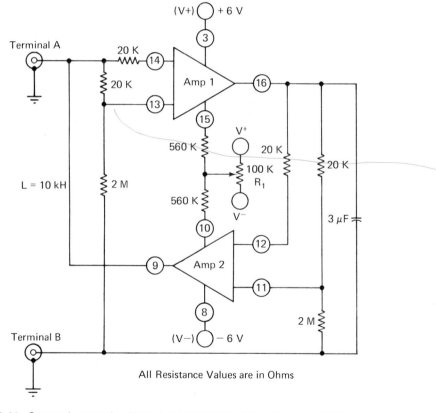

Fig. 3-32. Gyrator in an active filter circuit using two OTAs (Courtesy RCA).

In the simplest form, an OTA can be connected as a conventional amplifier, but with the I_{ABC} input connected to a voltage source through a variable resistance (which acts as the gain control). For a specified value of I_{ABC} (as set by the variable resistance), the output current of the OTA is equal to the product of transconductance and the input signal magnitude. The output voltage swing is the product of output current and the load resistance.

This gain control function can be applied to amplitude modulation of signals. Figure 3-33 shows a basic amplitude modulation system using an OTA as the modulator. In this circuit, the signal input is a voltage V_X at some carrier frequency, and the I_{ABC} input is a voltage V_M at some modulating frequency. The output signal current I_O is equal to transconductance (g_m) times V_X. The sign of the output signal is negative because the input signal is applied to the inverting input of the OTA.

The transconductance of the OTA is controlled by adjustment of I_{ABC}, as usual. However, in this circuit, the level of the unmodulated carrier output is established by a particular I_{ABC} through resistor R_M. Amplitude modulation of the carrier frequency occurs because variations of the voltage V_M force a change in the I_{ABC} supplied via resistor R_M. When V_M goes in the negative direction (toward the I_{ABC} terminal potential), the I_{ABC} decreases and reduces g_m of the OTA. When V_M goes positive, the I_{ABC} increases, resulting in an increase of the g_m.

For the particular OTA shown (an RCA CA3080A), the g_m is approx-

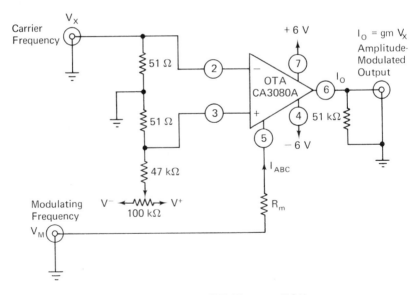

Fig. 3-33. Amplitude modulator using an OTA (Courtesy RCA).

imately equal to 19.2 x I_{ABC}, where g_m is in mmhos and I_{ABC} is in mA. In this case, I_{ABC} is approximately equal to:

$$\frac{V_M - (V^-)}{R_M} = I_{ABC}$$

$$(I_0) = -g_m V_X$$

$$g_m V_X = (19.2)\ (I_{ABC})(V_X)$$

$$I_0 = \frac{-19.2\ [V_M - (V^-)]\ V_X}{R_M}$$

$$I_0 = \frac{19.2\ (V_X)(V^-)}{R_M} - \frac{19.2\ (V_X)(V_M)}{R_M} \quad \text{(modulation equation)}$$

Note that there are two terms in the modulation equation. The first term represents the fixed carrier input, independent of V_M. The second term represents the modulation, which either adds to or subtracts from the first term. When V_M is equal to the V^- term, the output is reduced to zero.

In the preceding modulation equations, the term

$$(19.2)\ (V_X)\ \frac{V_{ABC}}{R_M}$$

involving the amplifier bias current terminal voltage V_{ABC} was neglected. This term was assumed to be small because V_{ABC} is small compared with V^- in the equation. If the amplifier bias current terminal is driven by a current-source (such as from the collector of a PNP transistor), the effect of V_{ABC} variation is eliminated. Instead, any variation is dependent upon the PNP transistor base-emitter junction characteristics. Figure 3-34 shows a method of driving the I_{ABC} input using a PNP transistor.

If an NPN transistor is added to the circuit of Fig. 3-34 as an emitter-follower to drive the PNP transistor, variations resulting from base-emitter characteristics of the PNP are considerably reduced because of the complementary nature of the NPN and PNP base-emitter junctions. Also, the temperature coefficients of the two base-emitter junctions tend to cancel one another. Figure 3-35 shows a configuration using one transistor (in an RCA CA3018A NPN transistor array) as an emitter-follower (with the three remaining transistors of the transistor array connected as a current-source for the emitter-followers).

Note that the circuits of Figs. 3-33 and 3-34 use ±6V supplies, whereas the circuit of Fig. 3-35 requires a ±15V supply. The 100K potentiometer shown in all three circuits is used to null the effects of amplifier input

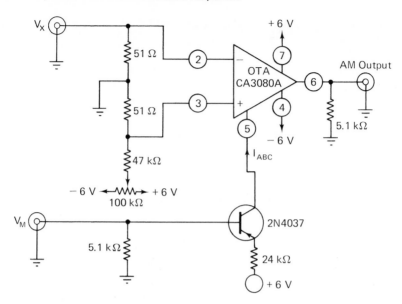

Fig. 3-34. Amplitude modulator using OTA controlled by PNP transistor (Courtesy RCA).

offset voltage (V_{io}). This potentiometer is used to set the output voltage symmetrically about zero (as described for conventional op-amps in Chapters 1 and 2). The OTA modulation method described here permits a range exceeding 1000-to-1 in gain, and thus provides modulation of the carrier signal input in excess of 99 percent.

3-5.6 OTA two-quadrant multiplier

Figure 3-36 shows an OTA used as a two-quadrant multiplier. Note that the circuit of Fig. 3-36a is essentially the same as for the modulator circuits described in Sec. 3-5.5. That is, when modulation is applied to the I_{ABC} input, and the carrier voltage is applied to the differential input, the waveform shown in Fig. 3-36b is obtained. However, in the circuit of Fig. 3-36, the input offset control (100K potentiometer) is adjusted to balance the circuit so that no modulation can occur at the output without a carrier input. The linearity of the modulator is indicated by the solid trace of the superimposed modulating frequency, as shown in Fig. 3-36b. The maximum depth of modulation (or percentage of modulation) is determined by the ratio of the peak input modulating voltage to V^-.

The two-quadrant multiplier characteristic of the circuit is seen if modulation and carrier are reversed (modulation to differential input,

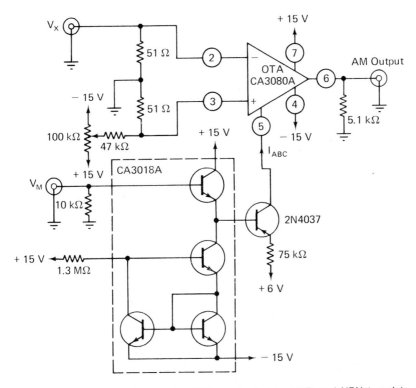

Fig. 3-35. Amplitude modulator using OTA controlled by PNP and NPN transistors (Courtesy RCA).

carrier to I_{ABC}), as shown in Fig. 3-36c. The polarity of the output must follow that of the differential input. Thus, the output is positive only during the first (or positive) half-cycle of the modulation, and is negative only during the second half-cycle. Note that both input and output signals are referenced to ground. The output signal is zero when either the differential input or I_{ABC} is zero.

3-5.7 OTA four-quadrant multipliers

OTA units can be used as four-quadrant multipliers. Two circuit configurations are possible. One circuit uses three identical OTA devices. The second circuit uses a single OTA.

Figure 3-37 shows a block diagram of a four-quadrant multiplier using the three OTA units of an RCA CA3060 package. (As discussed in Sec. 3-5.1, the CA3060 is an IC package that contains three identical OTA units.)

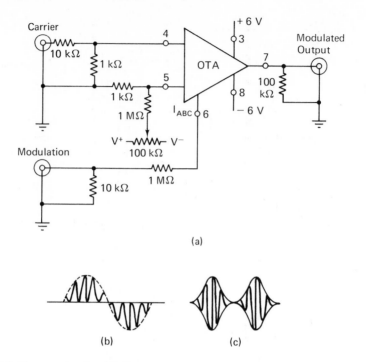

Fig. 3-36. Two-quadrant multiplier.

As shown in Fig. 3-37, amplifier 1 is connected as an inverting amplifier for the X-input signal. The output current of amplifier 1 is calculated as:

$$I_O(1) = -V_X \, g_m(1) \qquad \textbf{(Eq. 3-1)}$$

Amplifier 2 is connected as a noninverting amplifier so that:

$$I_O(2) = +V_X \, g_m(2) \qquad \textbf{(Eq. 3-2)}$$

Because the amplifier output impedances are high, the load current is the sum of the two output currents for an output voltage:

$$V_O = V_X \, R_L \, [g_m(2) - g_m(1)] \qquad \textbf{(Eq. 3-3)}$$

The transconductance is approximately proportional to the I_{ABC}. The I_{ABC} of amplifier 2 is proportional to the Y-input signal and is expressed as:

$$I_{ABC}(2) \approx \frac{(V^-) + V_Y}{R_2} \qquad \textbf{(Eq. 3-4)}$$

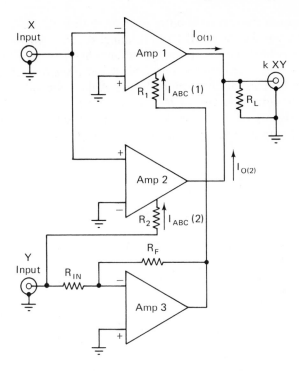

Fig. 3-37. Four-quadrant multiplier using OUAs (Courtesy RCA).

Hence,

$$g_m(2) \approx k\left[(V^-) + V_Y\right] \qquad \text{(Eq. 3-5)}$$

Bias for amplifier 1 is derived from the output of amplifier 3, which is connected as a unity-gain inverting amplifier. Thus, $I_{ABC}(1)$ varies inversely with V_Y. Hence,

$$g_m(1) \approx k\left[(V^-) - V_Y\right] \qquad \text{(Eq. 3-6)}$$

Combining equations 3, 5 and 6 yields:

$$V_O \approx V_X \cdot k \cdot R_L \left[[(V^-) + V_Y] -](V^-) - V_Y]\right] \text{ or}$$
$$V_O \approx 2\,K\,R_L\,V_X\,V_Y$$

Figure 3-38 shows a complete schematic for the four-quadrant multiplier, including all of the adjustment controls associated with differential input, and an adjustment for equalizing the gains of amplifiers 1 and 2.

Adjustment of the circuit is quite simple. With both the X and Y voltages at zero, connect terminal 10 to terminal 8. This procedure disables amplifier 2. Adjust the offset voltage of amplifier 1 to zero by means of R_1. Remove the short between terminals 8 and 10. Connect terminal 15 to terminal 8. This disables amplifier 1. Adjust the offset voltage of amplifier 2 to zero by means of R_2.

With ac signals on both the X and Y inputs, adjust R_3 and R_{11} for symmetrical output signals. Figure 3-39 shows the output waveform with

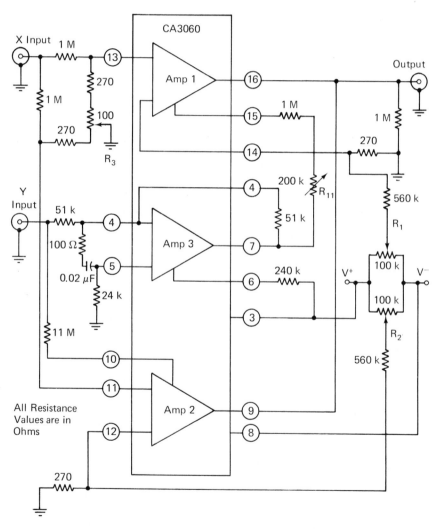

Fig. 3-38. Typical four-quadrant multiplier circuit using CA3060 OTA (Courtesy RCA).

the circuit adjusted. Fig. 3-39a shows suppressed carrier modulation of a 14 kHz carrier with a triangular wave. Figs. 3-39b and 3-39c, respectively, show the squaring of a triangular wave and a sine wave. Notice that in both cases the outputs are always positive, and return to zero after each cycle.

Four-quadrant multipliers using a single OTA — A single OTA can be used for many low-frequency, low-power, four-quadrant multiplier applications. The basic multiplier circuit shown in Fig. 3-40 using an RCA CA3080A is particularly useful in waveform generation, doubly balanced modulation and other signal processing applications, in portable equipment where low-power consumption is essential and accuracy requirements are moderate. The multiplier circuit of Fig. 3-40 is basically an extension of the gain-control function discussed in Sec. 3-5.5. Note that the single OTA version of the four-quadrant multiplier does not provide the accuracy of the system using three OTA units.

To obtain a four-quadrant multiplier, the first term of the modulation equation (which represents the fixed carrier) must be reduced to zero. In the circuit of Fig. 3-40, this is accomplished by placing a feedback resistor R between the output and the inverting input terminal of the OTA. The value of the feedback resistor R is equal to $1/g_m$.

The output current I_0 is equal to $g_m(-V_x)$ since the input is applied to the inverting terminal of the OTA. The output current due to the re-

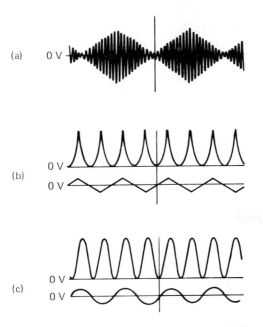

Fig. 3-39. Voltage waveforms of four-quadrant multiplier circuit (Courtesy RCA).

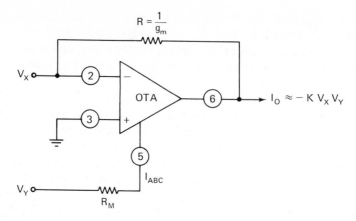

Fig. 3-40. Basic four-quadrant analog multiplier using an OTA.

sistor R is equal to V_X/R. Thus, the two signals (V_X and V_Y) cancel when R equals $1/g_m$.

For the particular OTA shown, the current is:

$$I_0 = \frac{-19.2\ V_X V_Y}{R_M}$$

The output signal for the single OTA configuration is a "current" that is best terminated by a short circuit. This condition can be satisfied by making the load resistance for the multiplier output very small. An alternate method for lowering the load resistance is to feed the output of the multiplier into a current-to-voltage converter (refer to Chapter 2). Figure 3-41 shows such a circuit.

The current "cancellation" in the feedback resistance (R of Fig. 3-40) is a direct function of the OTA differential amplifier linearity. If the OTA is operated on a nonlinear portion of its range, "cancellation" will not be complete, and the first term of the modulation equation may not be reduced to zero. To prevent this condition, operate the OTA at low signal levels. In general, nonlinear operation occurs when the OTA is overdriven. In the following examples, signal excursion is limited to ±10 mV to preserve linearity.

Figure 3-42 shows the schematic diagram of a basic multiplier, with adjustment controls added to give the circuit an accuracy of approximately ±7 percent of full-scale. There are only three adjustment controls: R_1 at the output compensates for slight variations in the current-transfer ratio of the current-mirrors (which would otherwise result in a symmetrical output about some current level other than zero); R_2 in the V_Y

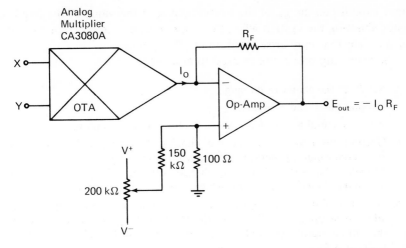

Fig. 3-41. OTA analog multiplier driving an op-amp that operates as a current-to-voltage converter.

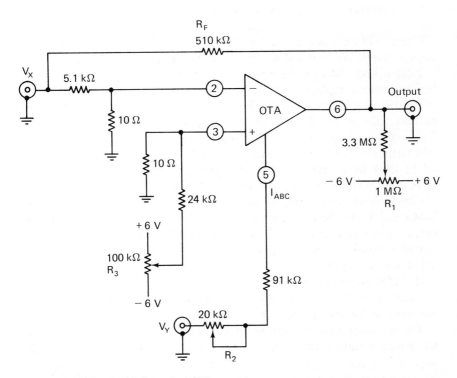

Fig. 3-42. Analog multiplier using OTA.

input establishes the g_m of the system equal to the value of the fixed resistor shunting the system when the Y-input is zero; and R_3 compensates for an error that may arise from input offset voltage.

The following procedure is used to adjust the circuit:

1. Set R_1 to the center of its range.
2. Ground the X and Y inputs.
3. Adjust R_3 until a zero volt reading is obtained at the output.
4. Ground the Y-input and apply a signal to the X-input through a low source-impedance generator. (It is essential that a low impedance source be used. This will minimize any change in the g_m balance for zero-point due to the Y-input bias current (which is an I_{ABC} of about 50 μA).)
5. Adjust R_2 until a reading of zero volts is obtained at the output. This adjustment establishes the g_m of the OTA at the proper level to cancel the output signal. The output current is diverted through the feedback resistance R_F.
6. Ground the X-input and apply a signal to the Y-input through a low source-impedance generator.
7. Adjust R_3 for an output of zero volts.

There will be some interaction among the adjustments and the procedure should be repeated for best circuit performance.

Figure 3-43 shows the schematic of an OTA multiplier circuit with a PNP transistor replacing the Y-input "current" resistor. The advantage of the Fig. 3-43 circuit is in the higher input resistance resulting from the current gain of the PNP transistor. The addition of another emitter-follower preceding the PNP transistor (similar to that shown in Fig. 3-35) will further increase the current gain, while markedly reducing the effect of the PNP base-emitter temperature-dependent characteristic (and possible input offset due to base-emitter voltage).

Figure 3-44 shows output signals of the Fig. 3-43 circuit. Figures 3-44a and 3-44b show outputs when the circuit is used as a suppressed-carrier generator. Figures 3-44c and 3-44d show outputs when the circuit is used in signal squaring (that is, in squaring sinewave and triangular-wave inputs).

If ±15V power supplies are used (as shown in Fig. 3-43), both inputs can accept ±10V input signals. Adjustment of the circuit in Fig. 3-43 is the same as for the circuit in Fig. 3-42.

The accuracy and stability of these multipliers are direct functions of the power supply voltage stability because the Y-input is referred to the negative supply voltage. Tracking of the positive and negative supply is also important because the balance adjustments for both the offset voltage and output current are also referenced to these supplies.

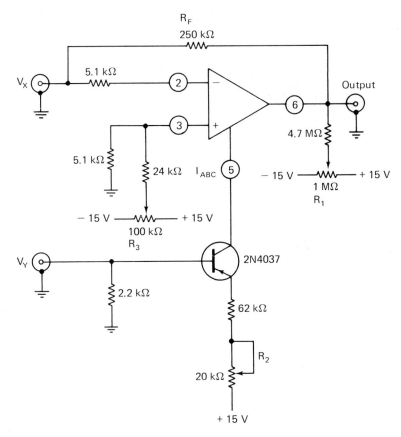

Fig. 3-43. Analog multiplier using OTA controlled by a PNP transistor.

3-5.8 OTA decoder-multiplexer

A simple but effective system for multiplexing and decoding can be assembled using OTAs as the basic elements. The complete circuit is shown in Fig. 3-45. Although only two channels are shown, more channels can be added. Figure 3-46 shows waveforms of circuit operation.

In the multiplexer, an IC flip-flop is used to trigger two OTAs, one OTA for each channel. The decoder consists of an OTA and MOSFET used as a sample-and-hold circuit (refer to Sec. 3-5.3) that is driven by another OTA used as a one-shot multivibrator. The OTA one-shot multi-vibrator introduces a 10 μS delay in the decoder to ensure that the sample-and-hold circuit can sample only after the input has settled. Thus, the

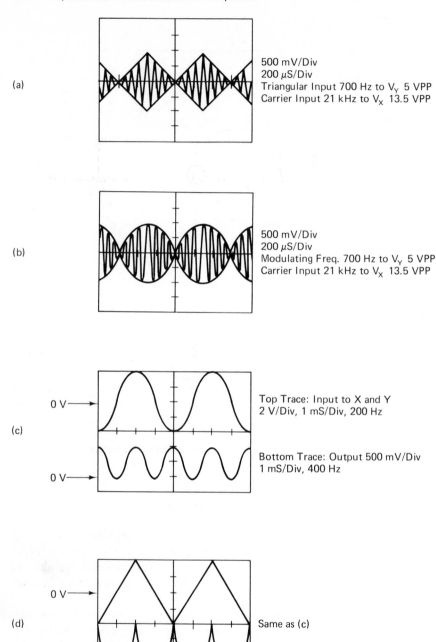

(a)

500 mV/Div
200 μS/Div
Triangular Input 700 Hz to V_Y 5 VPP
Carrier Input 21 kHz to V_X 13.5 VPP

(b)

500 mV/Div
200 μS/Div
Modulating Freq. 700 Hz to V_Y 5 VPP
Carrier Input 21 kHz to V_X 13.5 VPP

(c)

0 V ⟶

0 V ⟶

Top Trace: Input to X and Y
2 V/Div, 1 mS/Div, 200 Hz

Bottom Trace: Output 500 mV/Div
1 mS/Div, 400 Hz

(d)

0 V ⟶

0 V ⟶

Same as (c)

Fig. 3-44. Waveforms of analog multiplier.

trailing edge of the OTA one-shot output signal is used to sample the input at the sample-and-hold circuit for approximately 1 μS. Note that either the Q or $\overline{Q}$ output from the flip-flop may be used to trigger the 10 μS one-shot to decode a signal.

3-5.9 OTA with high current output stages

It is possible to combine an OTA with an MOS device to produce high-gain, high-current circuits. For example, the sample-and-hold circuit of Fig. 3-27 (Sec. 3-5.3) combines an OTA with a MOSFET. The resultant voltage gain is about 100 dB. The actual voltage gain of the overall circuit is equal to the product of OTA g_m and output resistance (which is typically 142,000 or 103 dB). Thus, the overall gain is set by the OTA characteristics. However, the output voltage-swing and current-swing are set by the MOSFET characteristics (and the source-terminal load).

Figure 3-47 shows an OTA combined with an MOS inverter/amplifier to form a simple, open-loop amplifier circuit. The MOS device shown is one-third of an RCA CD4007A COS/MOS inverter. (The term COS/MOS is the RCA designation for complementary-symmetry MOS devices. Such devices are discussed in the author's *Manual For MOS Users* [Reston Publishing Company, Reston, Va., 1974].) Each of the three inverter/amplifiers in the CD4007A has a typical voltage gain of 30 dB. This gain, combined with the typical 100 dB gain of the OTA, results in a total voltage gain of about 130 dB. (Note that OTA gain is affected by I_{ABC} which, in turn, is set by R_{ABC}. The rules for selecting R_{ABC} values are the same as described in Sec. 3-4.)

There are several circuit configurations available when an OTA is combined with MOS inverter/amplifiers. Some of these configurations provide additional gain, while others provide additional current capacity.

The circuit shown in Fig. 3-48 provides unity gain (with the same resistance values for R_F and R_I), and increased current capacity. The MOS inverter/amplifier (one-third of a CD4007A) can source or sink a current of about 6 mA.

The circuit shown in Fig. 3-49 provides a voltage gain of about 160 dB (10^8), and a source or sink current capacity of about 12 mA. The voltage gain results from open-loop operation. The OTA provides about 100 dB, MOS inverter/amplifier A provides about 30 dB and the parallel-connected MOS inverter/amplifiers B and C provide the remaining 30 dB. Since the outputs of B and C are connected in parallel, the normal source/sink current capacity of 6 mA is doubled to 12 mA.

The circuit of Fig. 3-50 provides unity gain (with the same resistance values for R_F and R_I), but with increased current capacity. The parallel-

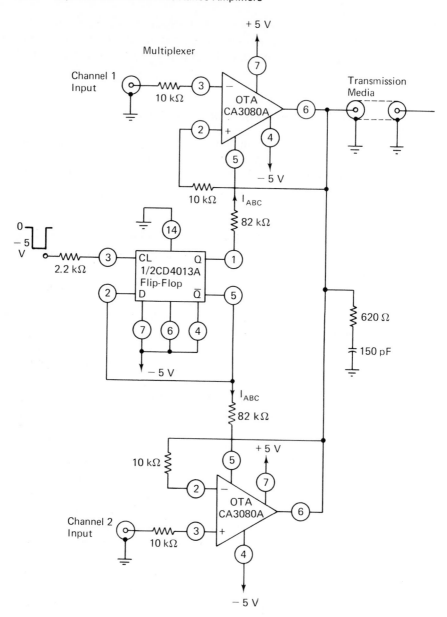

Fig. 3-45. Two-channel multiplexer and decoder using OTAs (Courtesy RCA).

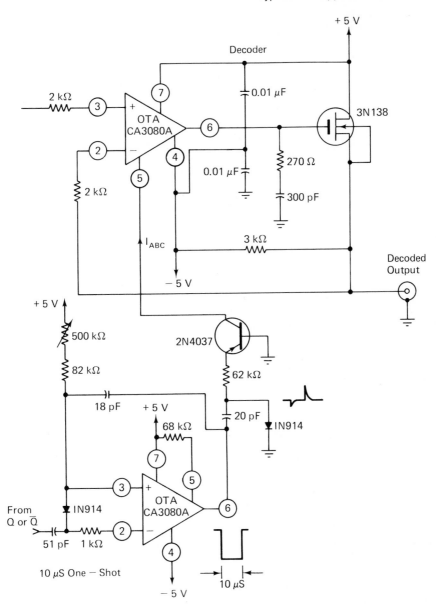

Fig. 3–45. (continued).

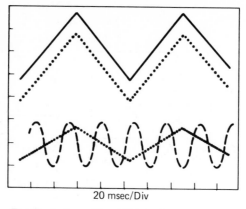

Top Trace: Input Signal (1 Volt/Div)
Center Trace: Recovered Output (1 Volt/Div)
Bottom Trace: Multiplexed Signals (2 Volts/Div)

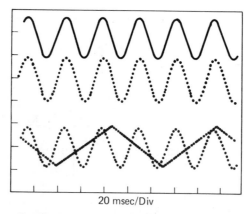

Top Trace: Input Signal (1 Volt/Div)
Center Trace: Recovered Output (1 Volt/Div)
Bottom Trace: Multiplexed Signals (1 Volt/Div)

Fig. 3-46. Waveforms showing operation of linear multiplexer/sample-and-hold diode circuitry (Courtesy RCA).

connected B and C outputs provide approximately 12 mA of sink or source current capacity.

3-5.10 OTA multistable circuits

An OTA can be set to draw very little standby current by proper adjustment of I_{ABC}. Likewise, MOS inverter/amplifiers draw very little current when used in switching applications. An MOS inverter/

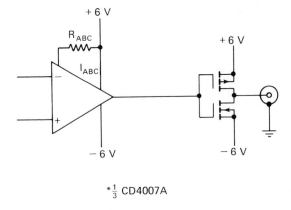

*⅓ CD4007A

*Additional current output can be
obtained when remaining two
amplifiers of the CD4007A are
connected in parallel with the
single stage shown.

Fig. 3-47. OTA combined with MOS to form open-loop amplifier circuit with high
current output.

amplifier switch draws current when changing states, but very little current when the output signal voltage swings either positive or negative. The low standby power capabilities of the OTA, when combined with the characteristics of the MOS inverter/amplifier, are ideally suited for use in precision multistable circuits.

Figure 3-51 shows an OTA combined with an MOS inverter/amplifier to form an astable (free-running) multivibrator. As shown by the equations, the frequency of operation is set by feedback resistance R and capacitor C, as well as resistors R_1 and R_2. Since resistors R_1 and R_2 also set output impedance of the circuit, a tradeoff may be necessary in selecting values of R and C. The value of R_{ABC} is set by the I_{ABC} requirements, as described in Sec. 3-4. For a multistable circuit, I_{ABC} is usually set so that the OTA draws minimum power, but enough so that the MOS inverter/amplifier is properly driven to provide the desired output voltage swing. Using the values shown, the output frequency is about 7.7 kHz, with an output voltage swing equal to $V+$ and $V-$.

Figure 3-52 shows an OTA combined with an MOS inverter/amplifier to form a monostable (one-shot) multivibrator. Generally, monostable multivibrators are used to produce output pulses of some specific time duration *(T)*, regardless of trigger input pulse duration and frequency. As shown by the equations, the time duration T of output pulses is set by feedback capacitance C and resistance R, as well as resistors R_1 and R_2.

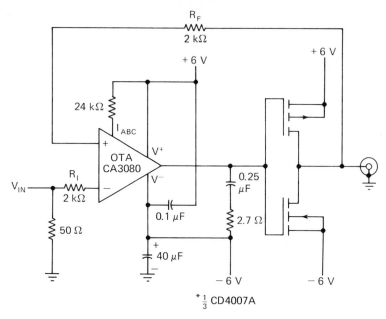

$$*\tfrac{1}{3} \text{ CD4007A}$$

*Additional current output can be
obtained when remaining two
amplifiers of the CD4007A are
connected in parallel with the
single stage shown.

Fig. 3-48. OTA combined with MOS to form unity gain amplifier with increased current
capacity.

The relationship of $V+$, $V-$ and V_D also affects output pulse duration T.
Note that V_D, the voltage across the input diode, is typically 0.5V. Again,
R_{ABC} is adjusted so that the OTA provides just enough output to drive
the MOS inverter/amplifier for the desired output voltage swing. Also,
it may be necessary to trade off the values of R_1, R_2, R and C, since R_1
and R_2 set the output impedance of the circuit.

Figure 3-53 shows an OTA combined with an MOS inverter/amplifier
to form a threshold detector. The threshold voltage point is set by R_1
and R_2, as well as the supply voltage. Standby power consumption is set
by I_{ABC}, which, in turn, is set by R_{ABC}.

The standby power consumption of the circuits shown in Figs. 3-51
through 3-53 is typically 6 mW. However, the standby power can be
made to operate in the micropower region by changes in the value of
R_{ABC}. Also, for greater current output from any of the circuits shown in
Figs. 3-51 through 3-53, the remaining MOS amplifier/inverters in the

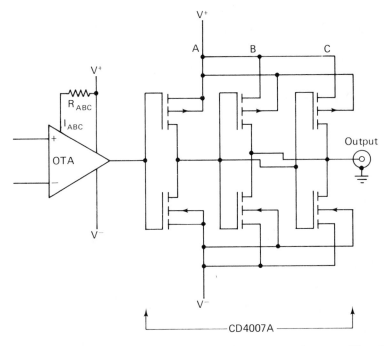

Fig. 3-49. OTA combined with three MOS stages to form open-loop amplifier circuit with high current output and increased gain.

RCA CD4007A can be connected in parallel with the single stage. Each of the three elements in the CD4007A will sink or source about 6 mA. Thus, with all three elements in parallel, the circuit should be able to sink or source about 18 mA.

3-5.11 OTA micropower comparator

Figure 3-54 shows an OTA combined with two MOS inverter/amplifiers to form a micropower comparator. Circuit output is proportional to the differential signal at the OTA inputs. If both inputs are at the same level, there will be no output. Either of the inputs (inverting or noninverting) can be adjusted to some reference level by means of a voltage divider network (such as shown in Fig. 3-26) if desired. Under these conditions, the output is proportional to the difference between the signal input level and the reference level.

The circuit is "on" only when a "strobe" or gate pulse is applied. The standby power consumption of the circuit is about 10 μW. When the circuit is strobed, the OTA consumes about 420 μW. Under these conditions the circuit responds to a differential input signal in about 6 μS. By decreasing the value of R_{ABC}, the circuit response time can be decreased to about 150 nS. However, the standby power consumption will rise to about 20 mW.

The differential input common-mode range of the circuit is approximately −1V to +10.5V. Voltage gain of the circuit is about 130 dB.

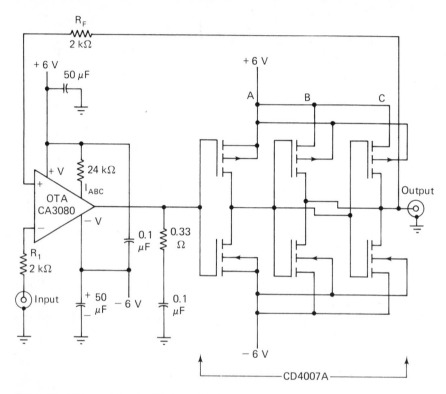

Fig. 3-50. OTA combined with three MOS stages for form unity gain amplifier with increased current capacity.

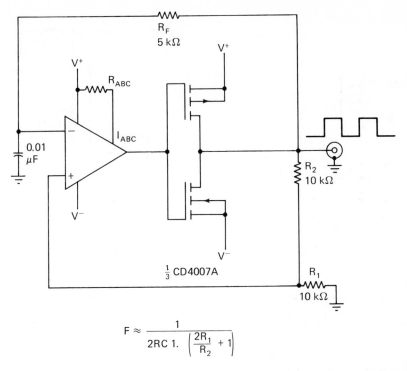

$$F \approx \frac{1}{2RC \cdot \left(\dfrac{2R_1}{R_2} + 1\right)}$$

Fig. 3-51. OTA combined with a MOS to form an astable (free-running) multivibrator.

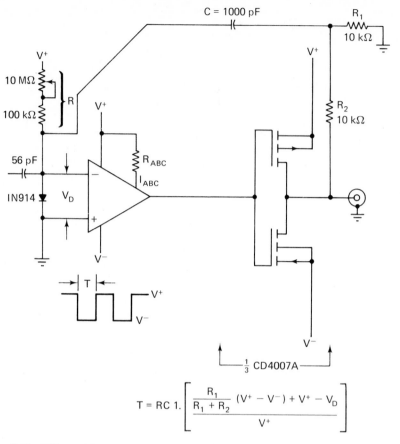

$$T = RC\ 1.\left[\dfrac{\dfrac{R_1}{R_1 + R_2}\ (V^+ - V^-) + V^+ - V_D}{V^+}\right]$$

Fig. 3-52. OTA combined with MOS to form a monostable (one-shot) multivibrator.

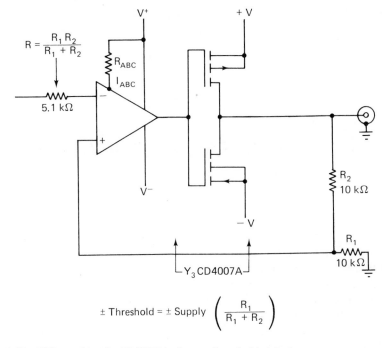

$$\pm\ \text{Threshold} = \pm\ \text{Supply}\ \left(\frac{R_1}{R_1 + R_2}\right)$$

Fig. 3-53. OTA combined with MOS to form a threshold detector.

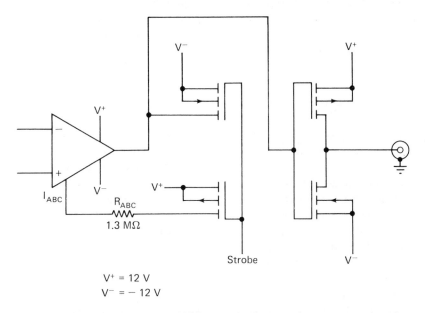

$V^+ = 12\ V$
$V^- = -\ 12\ V$

Fig. 3-54. OTA combined with two MOS stages to form a micropower comparator.

4. NONLINEAR APPLICATIONS

In this chapter, we shall discuss nonlinear applications for op-amps. These nonlinear applications include those cases in which inputs and outputs are essentially non-sinewave, or where the output is drastically modified by the op-amp. Also included are those circuits in which the op-amp is used to generate nonlinear signals (such as square-waves, pulses, integrated waves, etc.).

The format used here is essentially the same as found in Chapter 2. Unless otherwise stated, all of the design considerations for the basic op-amp described in Chapters 1 and 2 apply to each application covered here. Particular attention must be given to the information regarding op-amp power supplies and phase compensation discussed in Secs. 2-1 and 2-2, respectively.

4-1. PEAK DETECTOR

Figure 4-1 is the working schematic of an op-amp used as the major active element in a peak detector system. Where accuracy is required for peak detection, the conventional diode-capacitor detector is often inadequate because of changes in the diode's forward voltage drop (due to variations in the charging current and temperature). Ideally, if the forward drop of the diode can be made negligible, the peak value is the absolute value of the peak input, and will not be diminished by the diode-voltage drop (typically 0.5 to 0.7V for a silicon diode). This can be accomplished by means of an op-amp.

The circuit of Fig. 4-1 uses the base-collector junction of a transistor as the detecting diode. The transistor (acting as a diode) is contained within the feedback loop (between the op-amp output and the circuit output). This reduces the effective forward voltage drop of the diode by

172

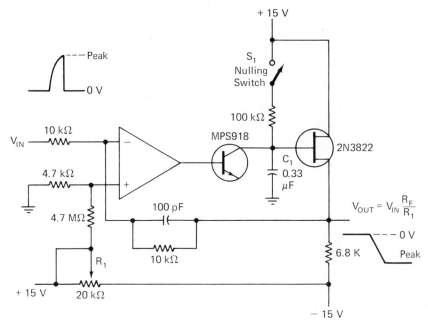

Fig. 4-1. Peak detector using op-amp.

an amount equal to the loop gain. For example, assume that the transistor has a base-emitter drop of 0.5V, that the open-loop gain of the op-amp is 1000, and the closed-loop gain is 1 (unity). Under these conditions, the loop gain is 1000, and the effective forward voltage drop is 0.0005V. Any transistor can be used provided the leakage is low (preferably 10 nA, or less, with +15V at the collector).

The storage time of the circuit in Fig. 4-1 is dependent upon leakage of the diode (transistor) and the FET, as well as the value of C_1. A larger value of C_1 will increase storage time.

Note that the values shown in Fig. 4-1 apply to a typical op-amp that requires a $V_{CC} - V_{EE}$ of 15V. These values can be used as a starting point for design.

In use, the output is adjusted to zero offset (zero volts output with no signal input) by closing the nulling switch S_1 and adjusting offset potentiometer R_1.

4-2. MULTIPLEX CIRCUIT

Figure 4-2 is the working schematic of two op-amps used as the active elements in a two-channel multiplex circuit. The purpose

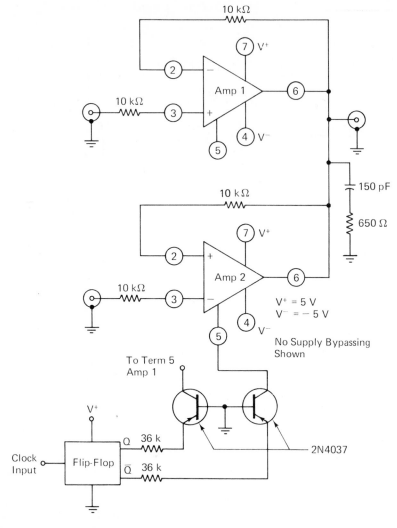

Fig. 4-2. Two op-amps used as active elements in two-channel multiplex circuit.

of such a circuit is to combine two or more signals (on a single line) using a time-sharing technique. In the circuit of Fig. 4-2, the two signals are introduced as the noninverting inputs of the op-amps. The circuit output appears at the parallel outputs of the op-amps. Each op-amp is alternately switched on and off by control pulses originating from a flip-flop, or other pulse source.

Note that conventional op-amps cannot be used in the circuit of Fig. 4-2. Instead, the op-amps described in Sec. 1-1.7 with external bias connections are required. The pulse signals from the flip-flop are amplified by two transistors connected as grounded-base (or common-base) amplifiers. The amplified clock signals are applied to the external bias connections of the op-amps. When one pulse is fully positive, the other pulse is fully negative. Thus, the op-amps are turned fully on and fully off on alternate cycles of the pulses. That is, op-amp 1 is on, and delivers the input signal to the output, when op-amp 2 is off, and vice versa.

Note that the op-amps are connected as unity gain (Sec. 2-4) amplifiers. Thus, the output is at essentially the same level as the inputs. If necessary, gain can be introduced by making the values of the feedback resistors larger than the input resistors, in the usual manner. The maximum circuit output voltage swing is limited by the values of $V+$ and $V-$. The capacitor and resistor at the circuit output are used to suppress transient voltages that may occur in the op-amps as a result of the pulses.

4-3. LINEAR STAIRCASE AND RAMP GENERATORS

Op-amps can be used as the active elements in linear ramp generator and linear staircase generator circuits. The staircase circuit (using an op-amp) is the more useful of the two since there are many other circuits capable of producing a linear ramp (and do not require the expense of an op-amp). However, design of a linear staircase generator using an op-amp is nearly identical to that of the ramp generator. For that reason, we shall discuss both types of circuits.

Figure 4-3 shows a linear ramp generator in which the noninverting input of an op-amp is grounded, and switch S_1 returns the output to the inverting input. When S_1 is closed, the op-amp is in the unity gain configuration, and the output is at ground (less any input offset voltage). When switch S_1 is opened, the output moves in the positive direction when the reference voltage V_{ref} is negative, or in the negative direction when V_{ref} is positive.

Because the output under closed-loop conditions (feedback through capacitor C) tries to maintain the input terminal at zero volts, the charging current to the capacitor C is constant at a rate of $dV/dt = I/C$, where: dV/dt is the increase (or decrease) of the ramp voltage for a given amount of time (such as one volt increase for each second after S_1 opens); C is the value of capacitor C (if C is expressed in μF and I is in amperes, then dV/dt is in volts per μS); I is V_{ref}/R (with V_{ref} in volts and R in ohms).

As an example, assume that V_{ref} is $+10$V, R is 100 ohms and C is 1 μF. Under these conditions, $I = 0.1$A ($+10$V/100 ohms) and the ramp decreases at a rate of one volt per 0.1 μS (0.1A/1 μF). (The ramp decreases since V_{ref} is given as a positive voltage.)

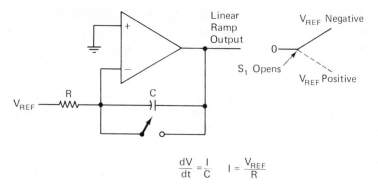

$$\frac{dV}{dt} = \frac{I}{C} \qquad I = \frac{V_{REF}}{R}$$

Fig. 4-3. Basic linear ramp generator using an op-amp.

Note that this equation for the linear ramp generator is accurate only if the *charging current is much greater than the input bias current* of the op-amp. As a guideline, make the charging current at least 100 times that of the input bias. If the bias is not small in relation to the charging current, the bias will offset the charging current (either add to or subtract from). This will tend to make the ramp nonlinear, particularly at the start or end of the ramp. In any event, the dV/dt equation will no longer be accurate.

The *maximum possible* height or amplitude of the linear ramp depends upon the supply voltages for the op-amps. That is, the maximum possible positive ramp is set by V_{CC}, whereas V_{EE} sets the maximum negative ramp. Assuming a given dV/dt (set by V_{ref}, R and C), the *actual height* of the ramp is set by the amount of time switch S_1 is open. For example, if dV/dt is one volt per μS, and S_1 is open for 10 μS, the ramp can increase from zero to 10V, provided the supply voltage is 10V or greater. In practical applications, the maximum ramp height is slightly less than the supply voltage.

Figure 4-4 shows the circuit for a *linear staircase generator using an op-amp*. Note that the circuit is similar to the ramp generator, except for the input. In the Fig. 4-4 circuit, charging resistance R is replaced by two diodes and capacitor C_1. Likewise, the fixed V_{ref} is replaced by pulses. In the circuit of Fig. 4-4, a pulse of amplitude E couples a charge Q to the op-amp input. The charge Q is equal to $C_1 (E - 2V_{ak})$, where $2V_{ak}$ is the forward voltage drop across the two diodes. Typically, $2V_{ak}$ is equal to about 1V, since the drop across each diode is about 0.5V.

When switch S_1 is open, capacitor C_2 is charged (by each pulse) in steps equal to $(E - 2V_{ak})(C_1/C_2)$. For example, assume that the pulses are 11V, C_1 is 500 pF and C_2 is 1 μF. Under these conditions, each step will be about 5 mV. This is shown as follows:

$$(11 - 1)(0.0005/1) = 0.005\text{V} = 5 \text{ mV}.$$

Note that the equations for the linear staircase generator are accurate only if the charging current is much greater than the input bias current of the op-amp. Thus, op-amps with minimum input bias are recommended for staircase generator circuits. Of more importance, the pulse amplitude must be much greater than the voltage/temperature variations of the diodes. If the diode voltage variations are not small in relation to the charging pulses, the voltage/temperature variations will add to (or subtract from) the pulse amplitudes, and produce uneven staircase steps.

The maximum possible height of the staircase is set by the op-amp supply voltages. The actual height of the staircase is set by the amount of time switch S_1 is open (assuming an infinite number of pulses and steps).

In practical ramp and staircase circuits, the switch S_1 is replaced with an electronic switch. Figure 4-5 shows a staircase generator circuit controlled by an FET switch. This circuit is used as part of a digital voltmeter. Clock pulses are applied to the FET gates to control operation of the staircase circuit. When the clock pulses are "on," the drain-source resistance of the FET is reduced to zero, and capacitor C_3 is shorted, or "closed." When the clock pulses are "off," the drain-source resistance rises to several hundred megohms, removing the "short" from C_3. With C_3 in the feedback circuit, the staircase generator produces steps in response to each signal pulse applied through C_1 and C_2. The generator will continue to produce steps until C_3 is again shorted by the FET in response to "off" clock pulses.

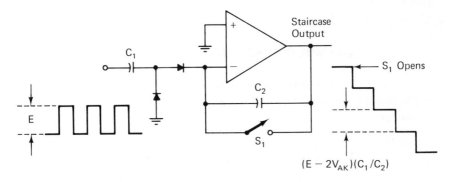

Fig. 4-4. Basic linear staircase generator using an op-amp.

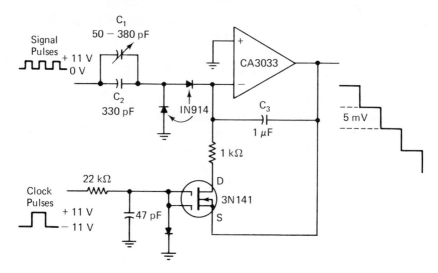

Fig. 4-5. Staircase generator circuit controlled by an FET electronic switch.

Note that capacitor C_1 is adjustable so that the staircase output can be set to exactly 5 mV steps. The signal pulses at the input are fixed at approximately 11V. This is reduced to about 10V by the drop across diodes D_1 and D_2. In theory, when C_1 is adjusted so that the parallel capacitance of C_1 and C_2 is 500 pF, the ratio of input capacitance to output capacitance is 0.0005 to 1, and the 10V input produces 5 mV steps at the output. However, because the diodes do not always have the same voltage drop, and the op-amp input bias is not always the same (due to temperature variations, etc.), the value of $C_1 - C_2$ is not always precisely 500 pF. Instead, C_1 is simply adjusted until the output steps are precisely 5 mV.

4-4. MULTIVIBRATORS

An op-amp can be used as the active element in various multivibrator circuits. The following paragraphs describe three typical examples.

4-4.1 Bistable multivibrator

Figure 4-6 is the basic diagram of a bistable multivibrator using an op-amp as the active element. The circuit shown is essentially a bistable *inverter,* and is so designated by some users. The circuit is also very similar to the op-amp comparator circuits discussed in Sec. 4-5.

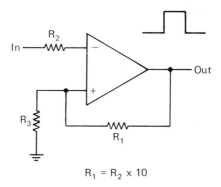

$$R_1 = R_2 \times 10$$

$$R_2 = R_3 = \text{Impedance of Signal Source}$$

Fig. 4-6. Bistable multivibrator (bistable inverter) using an op-amp.

No matter what term is used, the circuit functions to square the input signal. The output is a square or rectangular wave, the amplitude of which is set by the V_{CC} and V_{EE} voltages. For example, if V_{CC} and V_{EE} are +10V and −10V, respectively, the maximum output is approximately 20V (slightly less). The duration of the output waveform is set by the duration of the input signal and, to a lesser extent, by the ratio of R_1/R_2.

The input signal is applied to the inverting input. The output is fed back to the noninverting input through feedback resistors. This arrangement causes the op-amp to be driven into full saturation, or full cutoff, by input signals. In one condition the output is equal to V_{CC} (or slightly less). In the opposite condition the output is equal to V_{EE}. The feedback to the noninverting input causes the op-amp to make the transition rapidly, thus producing a squarewave or rectangular output.

Design of the circuit is relatively simple, so no design example will be given. As shown by the equations, $R_3 = R_2$, and both are set by the impedance of the signal source. The value of R_1 sets gain of the circuit. The value of R_1 is selected to provide enough gain to drive the op-amp into the full-on and full-off conditions. The amount of gain required is set by amplitude of the input signal. As a first trial value, make R_1 10 times larger than R_2 for a gain of 10. If the input signal does not drive the op-amp into full-on and full-off (that is, if the output is not square or rectangular), increase the value of R_1 until the output is squared off.

4-4.2 Astable (free-running) multivibrator

Figure 4-7 is the basic diagram of an astable multivibrator using an op-amp as the active element. The circuit is also known as a free-running multivibrator since the output is a train of pulses (or square-

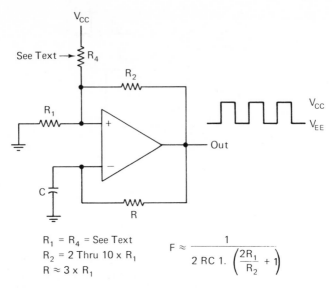

$$R_1 = R_4 = \text{See Text}$$
$$R_2 = 2 \text{ Thru } 10 \times R_1$$
$$R \approx 3 \times R_1$$

$$F \approx \frac{1}{2 \ RC \ 1. \ \left(\dfrac{2R_1}{R_2} + 1\right)}$$

Fig. 4-7. Astable (free-running) multivibrator using an op-amp.

waves) that are generated by the circuit. The circuit oscillates or generates the pulses without external signals or triggers.

In the astable multivibrator circuit, the output is fed back to both the inverting and noninverting inputs. The inverting input is returned to ground through capacitor C. The noninverting input is connected to ground through resistor R_1. The frequency of the pulse output is set by the charge and discharge of capacitor C through resistor R. To a lesser extent, the values of R_1 and R_2 in the noninverting feedback loop affect output frequency. This relationship is shown by the equation in Fig. 4-7.

As in the case of the bistable multivibrator, the maximum output amplitude is set by V_{CC} and V_{EE}. The values of R_1 and R_2 are set by the amount of gain required to produce oscillation. For a typical IC op-amp, the value of R_2 is at least twice the value of R_1, and possibly 10 times the value of R_1. The values of R and C are not critical, except that their combined time constant is the primary factor in setting output frequency. As a guideline, make R equal to three times R_1.

In most cases, the input bias current across R_1 produces enough voltage to start oscillation. However, if the voltage is not sufficient, a fixed, positive voltage (from V_{CC}) can be introduced to the noninverting input as shown. As a guideline, let R_4 equal R_1. This will divide V_{CC} so that the fixed bias to the noninverting input will be one-half of V_{CC}. Such a value will be more than enough to start oscillation. When a fixed bias from V_{CC} is not used, make R_1 a value that will produce a voltage drop greater than the input offset voltage, or at least as large.

4-4.3 Monostable (one-shot) multivibrator

Figure 4-8 is the basic diagram of a monostable multivibrator using an op-amp as the active element. The circuit is also known as a one-shot multivibrator since there is one output for each input. The monostable multivibrator is similar to the bistable multivibrator (Sec. 4-4.1). However, design is far more critical. For this reason, Fig. 4-8 includes a timing diagram. It is essential that the user study the timing diagram, as well as the following design considerations, when using the

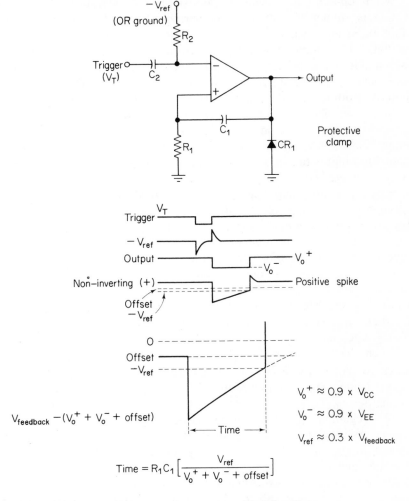

$$\text{Time} = R_1 C_1 \left[\frac{V_{ref}}{V_o{}^+ + V_o{}^- + \text{offset}} \right]$$

$R_2 C_2$ time constant $\approx 0.2 \times$ duration of negative trigger

Fig. 4-8. Monostable (one-shot) multivibrator using an op-amp.

circuit of Fig. 4-8. Although an op-amp can be readily used in the mono-stable circuit, failure to observe certain conditions will result in improper timing, and could damage the op-amp.

Basic theory. Like the bistable circuit, the function of a monostable multivibrator is to produce an output in response to an input trigger. Unlike the bistable circuit, the width or duration of a monostable output is set by the monostable circuit values, rather than by the trigger input. Thus, in a monostable multivibrator circuit, the output pulse can be "stretched" or set to a given width by selection of circuit values.

When op-amps are used as monostable multivibrators, the op-amp action is similar to that of a comparator (Sec. 4-5). One side of the differential input stage is set initially "on" whereas the remaining side is "off." The output stage is either saturated or cut off, depending upon which side of the differential input is involved. Therefore, the magnitude and polarity of the differential input voltage control the state of the output. Both magnitude and polarity can be set accordingly.

In the absence of an input trigger pulse, the noninverting input is near ground (except for the small voltage drop across R_1), and the inverting input is at the level of the reference voltage V_{ref}. Since these two levels constitute inputs to a differential input amplifier stage, the op-amp output is at a voltage level V_o^+, the maximum positive output voltage swing.

When positive-going trigger V_T is coupled to the inverting input through C_2, the input base is driven in a positive direction. When the inverting input becomes more positive than the noninverting input, the input differential stage reverses state and drives the op-amp output to level V_o^-, the maximum negative output voltage swing.

The output voltage is coupled through timing capacitor C_1 to the non-inverting input as a negative-going transient. Thus, the op-amp is effectively latched into the existing stage (output at V_o^-). Timing capacitor C_1 then charges toward ground through resistor R_1, as shown in the timing diagram.

When the voltage level at the noninverting input charges to a level equal to that at the inverting input (equal to $-V_{ref}$), differential action occurs again, and the output returns to level V_o^+. This completes the basic timing period.

The beginning of the timing period occurs as the result of a positive trigger. When the trailing edge of the trigger input is coupled to the in-verting input, the input is driven more negative, and the op-amp output remains unchanged (at V_o^-).

Design considerations. As shown by the equations of Fig. 4-8, am-plitude of the monostable output is approximately 90 percent of the supply voltage V_{CC}, V_{EE}. Thus, if V_{CC} and V_{EE} are 12V, the outputs are +10.8V and −10.8V. If the monostable is to be used in digital logic work (where logic levels are often in the order of 5V), V_{CC} and V_{EE} should be 6V.

At the beginning of the timing period (when the output switches to a negative state), a negative transient equal to the sum (V_0^+ and V_0^-) is fed back to the input. Precautions must be taken to ensure that the op-amp input is capable of accepting this input without damage (as is always the case when any input is applied to an op-amp).

For example, if V_0^+ and V_0^- are 10.8V, the noninverting input will receive about 21.6V at the start of the timing period. If this value is too high for the op-amp, protective clamp diode CR_1 can be added as shown in Fig. 4-8. With CR_1 in the circuit, the op-amp input receives the sum of V_0^+ and the threshold voltage of CR_1 (typically less than about 0.7V). Thus, if V_0^+ is 10.8, the op-amp input receives about 11.5V maximum.

In addition to the negative transient, the op-amp has an offset voltage developed across R_1. This voltage is determined primarily by the input bias current. The offset voltage must be added to the sum of the output voltages to find the actual voltage at the noninverting input when the timing period starts (as shown by the timing diagram). This total voltage is designated as $V_{feedback}$.

Resistor R_2 is returned to a negative reference voltage, rather than to ground. This reference voltage ensures complete cutoff of the op-amp during initial circuit conditions (before the positive trigger is applied). The reference voltage also controls the period of the output pulse. For example, as shown by the timing diagram, the output pulse period is increased if the reference voltage is decreased, and vice versa. The op-amp switches back to the initial state (end of the output pulse) when C_1 discharges to the level of the reference voltage.

The limits of the reference voltage are set primarily by the trigger voltage and the offset voltage across R_1. The negative reference voltage must be larger than the offset voltage (to ensure cutoff), but smaller than the trigger (to ensure that the trigger will switch states of the op-amp). As a secondary consideration, V_{ref} must be less than $V_{feedback}$. Typically, V_{ref} should be no greater than about one-third of $V_{feedback}$.

The period of the output pulse is determined by the RC time constant of R_1C_1, by the ratio of V_{ref} to $V_{feedback}$ and by constants of the op-amp. In practical use, trial values of R and C must be selected, and the actual period of the op-amp measured on an oscilloscope. However, the graph of Fig. 4-9 provides an approximate constant that can be applied to a typical op-amp. The curve shows a fixed $V_{feedback}$ of 6 to 7V versus a V_{ref} that varies between 0.5 and 2.5V. To find the approximate period of the output pulse with the graph, multiply the R_1C_1 time constant by the graph constant.

For example, assume a V_{ref} of -1V and an R_1C_1 time of 300 mS. The

−1V line (vertical) intersects the curve at a constant of about 1.75. Under these conditions, the period of the output pulse is 525 mS (300 × 1.75).

Keep in mind that the graph of Fig. 4-9 provides approximate values that will vary from op-amp to op-amp. Actual values must be determined by test. Also note that the graph does not go below about 0.5V (which is an aribtrary figure for the offset voltage across R_1) or above 2.5V (which is greater than one-third of $V_{feedback}$ and possibly greater than a typical trigger voltage).

The input R_2C_2 circuit must be capable of charging and discharging at a rate faster than the input. If not, it is possible that a slow or incomplete discharge of C_2 will affect the output timing, and could affect circuit operation. As a guideline, the duration of the negative trigger should be five times that of the R_2C_2 time constant. Often, R_2 must be chosen to match an input impedance.

In some uses, it is necessary that the output pulse not go to or below the zero level. This can be accomplished by means of the circuit shown in Fig. 4-10a. With this circuit the output pulse will not go below the value of V^+. Of course, V^+ must not be greater than V_o^+. Likewise, the total output swing is equal to the difference between V_o^+ and V^+.

In other uses, it is necessary that the output pulse go to the full positive and negative extremes (V_o^+ and V_o^-). If the combination of these voltages is above the same input level of the op-amp, a protective diode can be connected into the circuit as shown in Fig. 4-10b. The output then remains at full value, but $V_{feedback}$ is equal to V_o^+, plus the threshold value of the diode (about 0.7V).

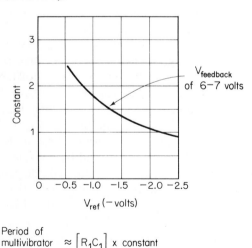

Period of
multivibrator $\approx \left[R_1C_1 \right]$ x constant
output pulse

Fig. 4-9. Graph for determining period of monostable multivibrator output pulse.

Note that the timing is affected by either circuit shown in Fig. 4-10. That is, the output pulse duration will not be the same as for the basic circuit of Fig. 4-8, even though identical circuit values are used. However, the values shown for Fig. 4-8 can be used as trial values for the Fig. 4-10 circuits, with actual timing determined by test.

Design example. Assume that the circuit of Fig. 4-8 is to provide an output pulse that goes from approximately $+8V$ to $-8V$, and then returns to $+8V$. The required output pulse duration is 300 mS. The input is a $-1V$ pulse with a duration of 30 mS, from a 50 ohm impedance source. The op-amp is rated for a maximum 10V input, and a maximum V_{CC}, V_{EE} of 12V.

The value of R_2 should be 50 ohms to match the input impedance of the pulse source.

The R_2C_2 time constant should be 0.2 times the input pulse duration, or 0.2×30 mS $= 6$ mS. With an R_2 of 50 ohms, and a constant of 6 mS, the value of C_2 is 6 mS/50 $= 120$ μF.

With the desired output of $\pm 8V$, both V_{CC} and V_{EE} must be about 9V, and the $V_{feedback}$ is about 16V. This is greater than the maximum 10V input. The feedback can be reduced, but with full output, by using the diode clamp modification shown in Fig. 4-10b. With this modification,

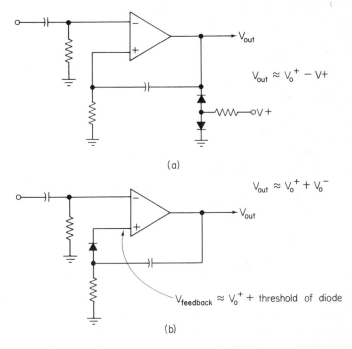

$$V_{out} \approx V_o^+ - V+$$

(a)

$$V_{out} \approx V_o^+ + V_o^-$$

$$V_{feedback} \approx V_o^+ + \text{threshold of diode}$$

(b)

Fig. 4-10. Circuit modifications to set output level of monostable multivibrator.

the total $V_{feedback}$ is then on the order of 9.2V (or less), allowing about 8V for V_o^+, 0.7V for the threshold of the diode (at the noninverting input), and 0.5V or less for the voltage drop across R_1. Thus, the feedback is below the 10V maximum of the op-amp, but the output is still V_o^+ and V_o^-.

The values of R_1 and C_1, as well as V_{ref}, must be selected to get the desired 300 mS output pulse duration. The value of V_{ref} must be greater than the offset voltage of 0.5V, but less than the trigger of -1V. Assume a value of 0.7V for V_{ref}. If the op-amp does not have enough gain to switch states with a V_{ref} of 0.7V, lower V_{ref} as necessary. In practical applications, the offset voltage across R_1 will probably be far less than 0.5V.

Using the time equation of Fig. 4-8, the ratio of V_{ref} (0.7V) to $V_o^+ + V_o^-$ + offset (about 16.5V) is approximately 0.4. Thus, R_1C_1 must have a time constant of 300 mS, when multiplied by 0.4. Divide 300 mS by 0.4 to find a constant of 750 mS. Any combination of R_1C_1 values can be used, provided that the product is equal to 750 mS.

From a practical standpoint, a larger value of R_1 (and a corresponding value of C_1) will keep the physical size of R_1C_1 small. However, the input bias current of the op-amp will produce a voltage drop across R_1, and a very large value of R_1 will produce a large drop. In turn, this large drop could affect the output pulse timing and circuit operation. As a guideline, use a value of R_1 that produces a drop (as a result of op-amp input bias current) equal to 10 percent (or less) of the offset drop (as a result of charging C_1). For example, with an assumed offset of 0.5V, the bias current drop should be 0.05V, or less. Assuming a 200 nA input bias, the value of R_1 should be 250K (0.05V/200 nA). With R_1 at 250K, and a required 750 mS, the value of C_1 should be 3 μF (750 mS/250K).

Keep in mind that the equations of Fig. 4-8 are subject to error when the modifications of Fig. 4-10 are used. Even without modification, the basic equations are approximations. Thus, in practical work, the completed circuit should be checked for proper output on an oscilloscope. If necessary, adjust voltage and/or component values to obtain the desired results. In summary, pulse duration is set by R_1, C_1 and V_{ref}. Pulse amplitude is set by V_{CC} and V_{EE}.

4-5. COMPARATORS

Figures 4-11 and 4-12 are the working schematics of op-amps used as comparators. A typical use for such comparators is in digital work where the circuit must produce a logic "1" or "0" output, depending on the polarity of a differential input signal (around some given reference level).

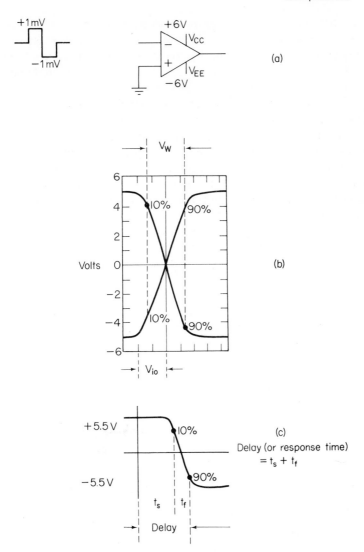

Fig. 4-11. Basic op-amp comparator characteristics.

Op-amps are essentially high-gain differential amplifiers (as discussed in Chapter 1). Without any circuit modification an op-amp can be used as a logic comparator. For example, in Fig. 4-11, if the noninverting input is connected to ground, and the inverting input varies between −1 mV and +1 mV, an op-amp with an open-loop gain of 10,000 will produce an output of −10V and +10V, respectively. If V_{CC} and V_{EE} are 10V, or less, the op-amp is driven into saturation by 1 mV inputs. This is shown by the voltage transfer characteristics of Fig. 4-11b.

Note that the edge of the output waveform is not perfectly vertical. Part of this condition is the result of the op-amp offset voltage (given as V_{io}), which must be overcome by the differential input signal. For example, if the offset is 1 mV, and a 1 mV differential is required to produce saturation, then there must be a 2 mV input.

There is also a delay of the signal through the op-amp. This delay or "response time" (shown as transition voltage or width, V_W, in Fig. 4-11b) is dependent upon op-amp characteristics, and cannot be altered.

Although some comparators are used with dc voltages or sinewaves, in digital work comparators are most often used to sense the difference in voltage between a pulse waveform and a fixed reference voltage. When working with pulses, switching times become important, rather than V_W. This is shown in Fig. 4-11c. The delay (or response time) is the sum of

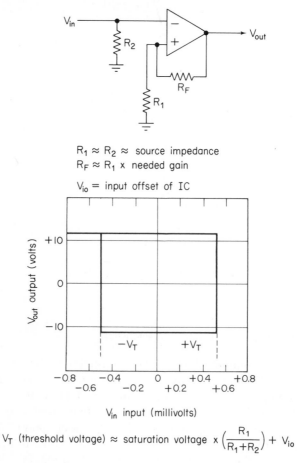

$R_1 \approx R_2 \approx$ source impedance

$R_F \approx R_1 \times$ needed gain

$V_{io} =$ input offset of IC

V_T (threshold voltage) $\approx$ saturation voltage $\times \left(\dfrac{R_1}{R_1 + R_2} \right) + V_{io}$

Fig. 4-12. Op-amp comparator using feedback to improve response time.

the storage time (t_s) and fall time (t_f). The storage time is essentially caused by saturation occuring in the latter stages of the op-amp. Fall time is essentially controlled by the parasitic capacitance in the op-amp input stages. Both factors are characteristics of the op-amp and cannot be changed.

There are two generally accepted methods to minimize delay (or to improve response time). One way is to overdrive the amplifier. For example, if a 1 mV differential produces saturation, then use 5 to 10 mV. Typically, overdrive should not exceed a factor of about 20. That is, if a 1 mV differential is needed, do not exceed 20 mV. Any further overdrive will probably not improve response time, but could result in malfunction.

The other method used to improve comparator response time is to use *positive feedback* as shown in Fig. 4-12. The positive feedback causes the circuit to switch into the desired state in a manner similar to that of the monostable multivibrator (Sec. 4-4.3). The feedback circuit helps drive the op-amp into saturation.

As shown by the equations of Fig. 4-12, the threshold voltage V_T (voltage differential required to switch states) becomes a function of maximum output voltage, times the ratio of feedback resistors, plus any offset voltage V_{io}. In effect, V_W is reduced to zero and is replaced by a more meaningful term V_T. (Of course, the delay time can never truly be zero since response time is also limited by the slew rate of the op-amp.)

Some sensitivity is lost when using the positive feedback of Fig. 4-12. However, this loss of sensitivity generally improves noise immunity which may be advantageous in certain applications.

Another problem with the circuit of Fig. 4-12 is that a voltage is developed across R_1 due to input bias current. The drop across R_1 represents a *fixed differential* that must be overcome by the input voltage. This problem can be minimized by the addition of R_2. The values of R_1 and R_2 are approximately the same, or slightly different so as to offset any unbalance in the differential input of the op-amp.

With the circuit shown in Fig. 4-12, R_2 becomes an impedance for the input. Thus, R_2 should be chosen to match the source impedance. Generally, a low value (200 ohms or less) for the source impedance (and R_2) is best. Since the circuit is designed to saturate, the value of the source impedance (and R_2) becomes a factor in determining the response time because of the RC time constant of the source impedance, and the input capacitance of the op-amp. A low value of source impedance will minimize this effect. Along with minimizing time delay, a small source impedance will also give a smaller input noise voltage, which is always desirable.

Another method of overcoming the offset problem is to apply a fixed compensating voltage to the positive input as shown in Fig. 4-13a. The compensating voltage must be of opposite polarity, but equal to the total offset at the input. For example, if the normal offset of the op-amp is $+1$ mV at the noninverting input, and there is a $+1$ mV offset (caused by a differential of voltage drops across R_1 and R_2), a -2 mV compensating voltage is required.

The circuits in Figs. 4-11 and 4-12 compare the input voltage with zero volts, but can be used to compare an input voltage with a reference voltage other than ground (zero). However, the reference voltage cannot exceed the *maximum common-mode range* of the op-amp. When the input must be compared with a reference voltage, rather than ground or zero volts, the fixed reference voltage is applied to the noninverting input as shown in Fig. 4-13b.

Typical circuits. Figure 4-14 shows an op-amp comparator used in a digital voltmeter. Note that regeneration is supplied through a resistor

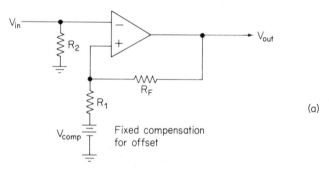

(a)

$$V_{comp} = V_{io} \text{ of IC + differential drops across } R_1 + R_2$$

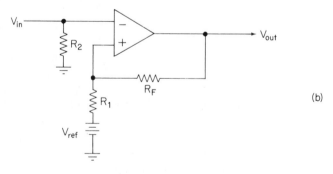

(b)

$$V_{ref} = \text{Desired reference + offset (if any)}$$
$$V_{ref} = < \text{Max. common mode input of IC}$$

Fig. 4-13. Methods to overcome input offset, and to set comparison about a fixed reference level.

and capacitor. This system isolates the input and output circuits in regard to direct current. The voltage applied to the inverting input varies between 0 and 7 mV. The voltage at the noninverting input is a series of 5 mV steps (supplied by a staircase generator such as shown in Fig. 4-5). The output of the comparator in Fig. 4-14 is a series of pulses (one for each step).

The two 0.001 μF capacitors at the op-amp inputs serve to filter any externally generated noise. Note that the input resistor values are higher than recommended. However, the op-amp shown has a nominal input bias current of 70 nA. This produces a nominal 0.7 mV drop across both 10K input resistors. Thus, the fixed differential is something less than 0.7 mV.

Figure 4-15 shows an op-amp comparator used in digital work. The input is a sweep voltage. The output is a series of pulses that occur when the sweep voltage goes above or below a given level, as set by the 5K potentiometer. The op-amp shown has an input bias current of about 225 nA, input offset voltage of about 0.5 mV, and a slew rate of about 40V/μS. Low input bias and offset voltage, as well as high slew rate, are required when an op-amp is used as a comparator.

The Zener diode connected to pin 5 of the op-amp limits the positive-going waveform at the output to about 2V below the Zener voltage (pin 5 is the collector of one output stage transistor). The silocon diode connected to the output limits the output negative excursion to protect the logic circuit.

The values of the parallel capacitor and resistor at the op-amp output

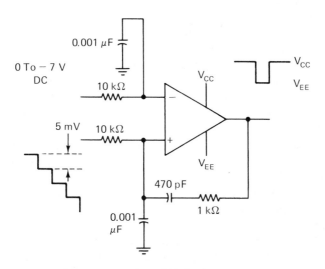

Fig. 4-14. Op-amp comparator used in a digital voltmeter.

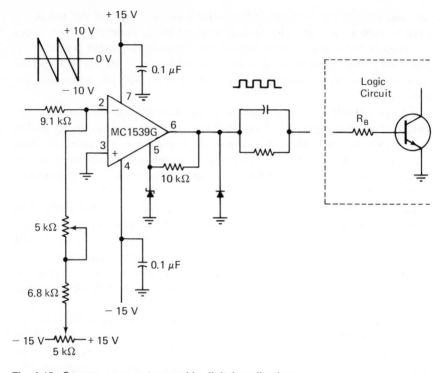

Fig. 4-15. Op-amp comparator used in digital applications.

are dependent upon the logic circuit, and not on the comparator. This parallel network has two functions. First, it matches the input of the logic circuit, eliminating any reduction in response time due to RC charging time. Second, some saturated logic circuits (MRTL for example) have a rather low value of input base resistance R_B (as low as 400 ohms), and the additional resistance in series with the comparator output will help minimize the output current overload problem. That is, the additional resistance will decrease output current drain on the op-amp.

The op-amp shown has protective diodes in the input. When the differential input exceeds about 0.75V, the diodes may begin to draw current. The effect is that when the differential input exceeds that amount needed to cause one of the input diodes to conduct, the input impedance of the comparator is reduced, causing a relatively heavier load to exist on the input signal source. In the circuit shown in Fig. 4-15, this is of no concern since only one comparison is being made. However, if many comparators are connected to a source, as is the case for an analog-to-digital converter with parallel banks of comparators, the excess current drain can cause error (if not malfunction).

4-5.1 Design example

Assume that the circuit of Fig. 4-12 is to compare an input against a fixed reference of $+1$V. The output is to switch states whenever the input goes above or below 1V by 0.5 mV. That is, any input above 1.0005 or below 0.9995V will cause the output to change to the opposite state. The available V_{CC} and V_{EE} are about 10V. The input source impedance is 50 ohms. The op-amp has an open-loop gain of about 50,000 or greater, and an input bias of 1000 nA. For simplicity, assume that there is no input offset, or that the op-amp has provisions for neutralizing the offset (as described in Chapters 1 and 2). The op-amp has a common-mode input maximum of 2V.

Since the source impedance is 50 ohms, make R_1 and R_2 50 ohms each. With a 1000 nA bias, the drop across R_2 is 0.05 mV. This is 10 percent of the 0.5 mV input, and can be ignored.

With V_{CC} and V_{EE} at 10V, the saturation is 10V. In practice, saturation will occur at some value slightly less than 10V.

With 10-V saturation, and an input of 0.5 mV, the needed gain is 20,000, well below the open-loop gain of 50,000. With a needed gain of 20,000, and an R_1 of 50 ohms, R_F must be 1 M ($50 \times 20,000$).

Using the equation of Fig. 4-12, the threshold voltage will work out to slightly less than 0.5 mV (about 0.4999 mV).

Since the comparator circuit is to operate about a $+1$V point, $+1$V must be applied to R_1 as shown in Fig. 4-13. The $+1$V is well below the common-mode range of the op-amp (2V). However, care should be taken that the input to be compared does not go below -1V, or above $+3$V. Otherwise, the common-mode range will be exceeded.

4-6. LOG AND ANTI-LOG AMPLIFIERS

Figure 4-16 is the working schematic of an op-amp used as a log (logarithmic) amplifier. A log amplifier is nonlinear so that a large input variation produces only a small output variation. This is shown by the curve of Fig. 4-17 where the input varies from 1 mV to 100V, but produces an output variation from about 350 mV to 640 mV (an approximate 300 mV output swing). Note that the output appears as a straight line on Fig. 4-17 since the horizontal lines represent logarithmic variations (five decades in this case).

The amplifier of Fig. 4-16 compresses the five decades of information into a small output swing. One circuit requiring such a log amplifier is a display that reads out data that span many orders of magnitude in a

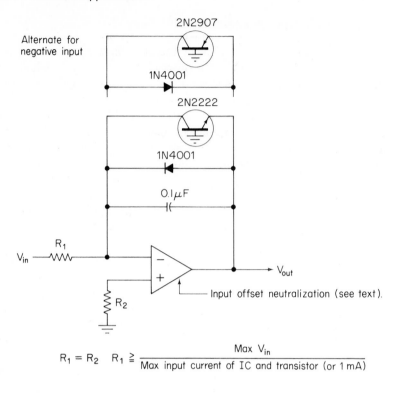

$$R_1 = R_2 \quad R_1 \geq \frac{\text{Max } V_{in}}{\text{Max input current of IC and transistor (or 1 mA)}}$$

$$R_1 \leq \frac{\text{Min } V_{in}}{\text{Input bias current of IC}}$$

Fig. 4-16. Logarithmic amplifier using op-amp.

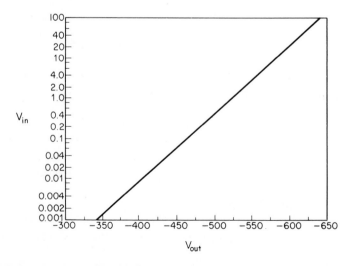

Fig. 4-17. Input versus output response of logarithmic amplifier.

single range. An NPN transistor is used when the input is positive. A negative input requires a PNP transistor, as shown in Fig. 4-16.

The circuit uses the base-emitter junction of a transistor to provide the logarithmic response. The principle of operation relies on the relationship of the base-emitter voltage and collector current of the transistor. All transistors do not exhibit good logarithmic characteristics, so care must be taken in selecting the proper transistor. Likewise, since the transistor junction is temperature sensitive, the amplifier response will also be subject to temperature variations.

Capacitor C_1 across the feedback transistor is necessary to reduce the ac gain (and thus reduce noise pickup). Use a value of 0.1 μF for C_1 for frequencies up to about 10 MHz. The diode CR_1 protects the transistor against excessive reverse base-emitter voltage should the polarity of the input voltage be reversed accidentally.

A log amplifier generally requires an op-amp that has provisions for neutralizing input offset voltage (as discussed in Chapters 1 and 2). The effects of offset voltage are noticed at the low limit of the input range. This is shown in Fig. 4-18.

If an op-amp without input offset provisions must be used, it is possible to provide offset neutralization with the circuit of Fig. 4-19. This circuit uses the available values of V_{CC} and V_{EE}, and provides both coarse and fine neutralization of the input offset voltage.

The circuit of Fig. 4-19 also provides for neutralization or compensation of the input bias current. As a rule, an op-amp with the lowest possible input bias current is best for log amplifiers. This permits operation with log input signal voltages.

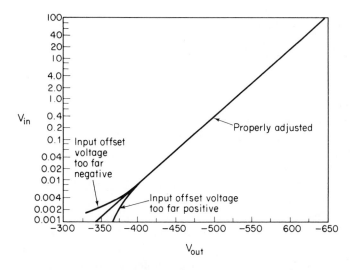

Fig. 4-18. Effects of improper offset adjustment on logarithmic amplifier.

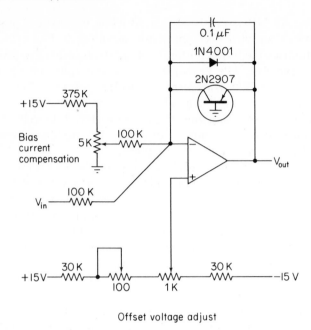

Offset voltage adjust

Fig. 4-19. Alternate method to provide offset neutralization (for op-amps without input offset provisions).

Referring back to the circuit in Fig. 4-16, the values of R_1 and R_2 are selected on the basis of input bias current, input voltage extremes and the current limits of the transistor and op-amp, as shown by the equations. When the signal input is at its high limit, there is a large voltage drop across resistor R_1 (almost the full value of the input signal). If a low value of R_1 is used, there is a large current flow through R_1. This current flow could exceed the input current limits of the op-amp and the transistor. In the absence of any specific datasheet information, use a value of 1 mA for maximum input current, when calculating the value of R_1. A large value of R_1 produces a corresponding voltage drop due to the input bias current. This voltage drop must be less than the low limit of the input voltage.

In the circuit of Fig. 4-19, the values of the bias and offset compensating networks are based on the limits of V_{CC} and V_{EE}. The current through the compensating networks should be less than 1 mA, and preferably in the order of 0.5 mA. The value of R_1 should be selected on the same basis as R_1 in Fig. 4-16. The value of R_2 should be equal to R_1.

Figure 4-20 is the working schematic of an op-amp used as an *anti-log amplifier*. The anti-log circuit is the complement of a log amplifier (the transistor is in the input circuit rather than as the feedback element). In an anti-log circuit, larger input voltages produce progressively higher amplifier gains. Because the input is applied across the transistor base-

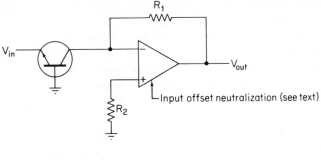

$R_1 = R_2 =$ See text

Fig. 4-20. Anti-logarithmic amplifier using op-amp.

emitter junction, input voltages are generally 1V or less. The maximum circuit output voltage is limited by the maximum voltage swing of the op-amp (which is set by V_{CC} and V_{EE}). The values of R_1 and R_2 should be selected on the same basis as R_1 and R_2 in Fig. 4-16.

4-6.1 Design example

Assume that the circuit of Fig. 4-16 is to provide a logarithmic output response similar to that shown in Fig. 4-17. The input voltage varies five decades, from 1 mV to 100V. Also assume that the op-amp has provisions for neutralizing any input offset voltage, that the input bias is 8 nA and that the maximum input current is 1 mA.

With a maximum input voltage of 100V, and a maximum input current of 1 mA, R_1 must be 100K or larger (100 V/ 1 mA).

With a minimum input voltage of 1 mV, and an input bias current of 8 nA, R_1 must be 125K or smaller (1 mV/8 nA). Use a trial value of 100K for both R_1 and R_2.

The curve of Fig. 4-17 is the "ideal" rather than the practical curve of a log amplifier using an op-amp. In use, the circuit characteristics must be measured and plotted on a log graph. It is not practical to predict exact output for a given input, since the final characteristics are directly dependent upon the log characteristics of the transistor used.

4-7. ENVELOPE DETECTOR

Figure 4-21 is the working schematic of two op-amps used as an envelope detector. The purpose of the circuit is to provide a visual indication (panel lamp) when an input voltage goes above or below two given reference voltages. The two op-amps operate essentially as comparators (Sec. 4-5). The output of the two op-amps is converted to a panel lamp indication by a logic circuit.

Both op-amps are connected for positive feedback so that they will switch states (similar to a comparator or multivibrator), rather than amplify any input differential as a linear output. A small differential input will drive the op-amps into saturation. However, the output of both op-amps is clamped to ground by CR_1 and CR_2. The negative (or low) output of each op-amp will not go below the normal voltage drop of CR_1 and CR_2 (typically 0.7V for silicon diodes). On the positive saturation, the op-amp output is about 90 percent of V_{CC}. Thus, with a V_{CC} of 6V, the op-amp outputs switch from about +5.4V to −0.7V. As output load increases, the +5.4V output may drop, but should remain above +5.0V.

Operation of the circuit can best be understood by reference to the curve of Fig. 4-22, and an example. Assume that the low reference V_1 is 2.5V, the high reference V_2 is 3.5V, V_{CC} is 6V, and the logic gates require 0V (or less) for a low (or false or "0") input, and +5V (or more) for a high (or true or "1") input. Now assume that the input signal varies between 2V, 3V and 4V, in turn.

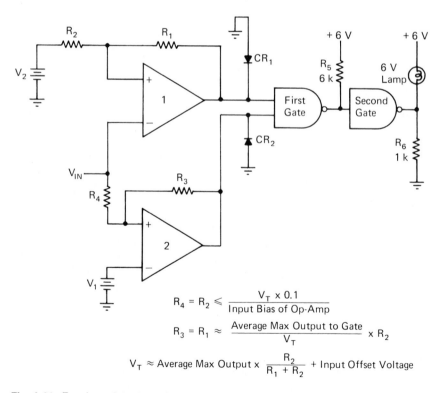

$$R_4 = R_2 \leqslant \frac{V_T \times 0.1}{\text{Input Bias of Op-Amp}}$$

$$R_3 = R_1 \approx \frac{\text{Average Max Output to Gate}}{V_T} \times R_2$$

$$V_T \approx \text{Average Max Output} \times \frac{R_2}{R_1 + R_2} + \text{Input Offset Voltage}$$

Fig. 4-21. Envelope detector using two op-amps.

With a 2V input signal, the input to op-amp 1 is low (2V is less than 3.5V), and the op-amp assumes the corresponding state. Since the signal input to op-amp 1 is at the inverting input, the output of op-amp 1 is high (saturated at about 5V). This represents a high input to the first logic NAND gate. With the same 2V signal, the input to op-amp 2 is low (2V is less than 2.5V) and, since the signal is at the noninverting input, the output of op-amp 2 is low to the first NAND gate.

A high and low input at a NAND gate produces a high output, which is inverted to a low by the second NAND gate. In effect, this shorts out R_6 (drops both sides of R_6 to ground or 0V), and places the full 6V across the lamp. Thus, the lamp turns on when the signal input (2V) is below the low reference (2.5V).

With a 4V input signal, the same action occurs, but the op-amp outputs transpose states. That is, the op-amp 1 output is low (−0.7V), the op-amp 2 is high (+5V), and the lamp turns on to indicate that the input has gone above the high reference.

With a 3V input signal, the input to op-amp 1 is low, and the output is high. The input to op-amp 2 is high, as is the output. Two high inputs to a NAND gate produce a low output. The low is inverted by the second NAND gate to a high, making both sides of the lamp about 6V. Thus, the lamp turns off when the signal input (3V) is above the low reference (2.5V) but below the high reference (3.5V).

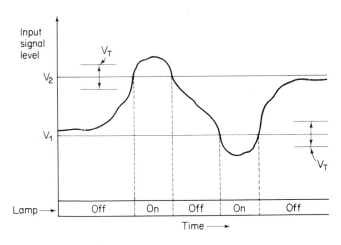

V_1 = low reference
V_2 = high reference
V_T = transition voltage (see text)

Fig. 4-22. Relationship of reference voltages, lamp indications and input voltage on op-amp envelope detector.

As shown by the curve of Fig. 4-22, there is a *transition voltage V_T* about both the high and low reference points. (This transition voltage is also known as *voltage spread, dead band, lag, hysteresis* and *threshold voltage*.) No matter what term is used, the effect is noticed when the signal voltage crosses the reference points. For example, if the signal is moving from 3V to 4V, the lamp will remain off until the signal has gone slightly higher than 3.5V. Likewise, when the signal goes from 4V to 3V, the lamp stays on until the signal is slightly below 3.5V.

Transition voltage V_T is a function of the maximum output voltage, the ratio of feedback resistors and any input offset voltage. However, the maximum output voltage for the op-amps of Fig. 4-21 is not the same for both states, so an average maximum output must be used. For example, if the op-amps go from -0.7V to $+5.4$V, the average maximum output is about 3V ($-0.7 + 5.4 = 6.1$; $6.1/2 = 3.05$). A *larger feedback produces a smaller V_T*, and vice versa.

The values of R_2 and R_4 are chosen on the basis of input bias current and voltage drop. The drop produced by bias current should be about 10 percent (or less) of V_T, rather than 10 percent of the input signal voltage. This is because the op-amp must operate (change states) on the small differential between signal and reference. The values of R_1 and R_3 are chosen to provide the desired gain. In the absence of some specific desired value of V_T, use a gain of about 50. Resistors R_3 and R_4 should be equal to the value of R_1 and R_2, respectively.

4-7.1 Design example

Assume that the circuit in Fig. 4-21 is to provide a V_T of about 80 to 100 mV. V_{CC} is 6V, V_1 is 2.5V, V_2 is 3.5V, input offset voltage is 5 mV and input bias current is 200 nA.

With a V_T of 80 mV, and an input bias current of 200 nA, the value of R_2 should be 40K maximum (80 mV $\times$ 0.1 = 8 mV; 8 mV/200 nA = 40K).

With a V_{CC} of 6V, the maximum voltage on the positive saturation is about 5.4V (6 $\times$ 90 percent = 5.4). The maximum negative swing is limited to about -0.7V by diodes CR_1 and CR_2. The average maximum output voltage is about 3V ($+5.4$ to -0.7 to 6.1; one half of 6.1 is 3.05V).

With an average maximum output voltage of 3V, a V_T of 80 mV, and a R_2 of 40K, the value of R_1 is 1500K (3 V/80 mV = 37.5; 37.5 $\times$ 40K = 1500K). Resistor R_3 should also be 1500K.

Using the values of 3V for average maximum voltage, 40K for R_1, 1500K for R_2 and an input offset voltage of 5 mV, find the V_T as shown by the equation of Fig. 4-21:

$$3V \times \left(\frac{40}{1500 + 40}\right) + 5 \text{ mV} \approx 95 \text{ mV}$$

If actual test of the circuits proves that V_T is below the 80 mV value, or above the 100 mV value, change the values of R_1 and R_3. An increase in R_1 and R_3 will decrease V_T, and vice versa.

The values for resistors R_5 and R_6 are not critical. These values shown on Fig. 4-21 are for use with a 6V lamp, and on the assumption that the logic circuit operates from the same 6V source (V_{CC}) as the op-amps.

4-8. TRACK AND HOLD AMPLIFIER (SAMPLE AND HOLD AMPLIFIER)

Figure 4-23 is the working schematic of two op-amps used as a track-and-hold amplifier (also known as a sample-and-hold amplifier).

With this circuit, when voltages are applied to the gate inputs, the diode bridge conducts and the output voltage tracks the input voltage. That is, V_{out} equals V_{in} and follows any variations of V_{in}. However, the polarity of V_{out} is reversed from that of V_{in}. If it is necessary that V_{out} track V_{in} directly, an op-amp connected as a unity gain amplifier (Chapter 2) can be used at the output.

With conventional silicon diodes, which have a normal voltage drop of about 0.7V, the gate voltage must be about 1.5V, plus the maximum input voltage. When the gate voltage polarity is reversed from that shown in Fig. 4-23 by an external switching system, the diode bridge stops conducting.

Due to the charge on capacitor C_1, the value of the output voltage is equal to that of the input voltage, just prior to switching the bridge off. The output voltage remains at this value for a time period, the length of which is dependent upon the value of C_1, the diode leakage current, and the input bias current of the op-amp. Neither of the currents (bias or diode leakage) can be altered. However, the time period can be set to a given value by proper selection of capacitor C_1 value.

The value of R_1 is chosen on the basis of input bias current and voltage drop. Typically, the drop across R_1 should be about one-tenth (or less) the input signal voltage, with nominal input bias current. Since the usual purpose of the circuit is to track the input voltage, no gain is required. Thus, R_2 should be the same value as R_1, at least as the first trial value. However, if the circuit shows some loss during test (output voltage is less than input voltage) increase the value of R_2 as necessary to produce the desired output.

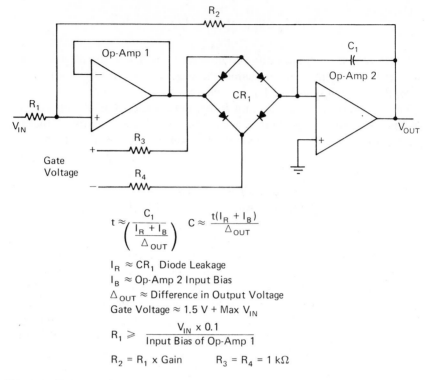

$$t \approx \cfrac{C_1}{\left(\cfrac{I_R + I_B}{\triangle_{OUT}}\right)} \qquad C \approx \frac{t(I_R + I_B)}{\triangle_{OUT}}$$

$I_R \approx CR_1$ Diode Leakage
$I_B \approx$ Op-Amp 2 Input Bias
$\triangle_{OUT} \approx$ Difference in Output Voltage
Gate Voltage ≈ 1.5 V + Max V_{IN}

$$R_1 \geq \frac{V_{IN} \times 0.1}{\text{Input Bias of Op-Amp 1}}$$

$R_2 = R_1 \times$ Gain $\qquad R_3 = R_4 = 1$ kΩ

Fig. 4-23. Track-and-hold (sample-and-hold) amplifier using op-amp.

The values of R_3 and R_4 are not critical. Use the 1K values for both resistors, assuming that standard silicon diodes are used, and that the gate voltage does not exceed about 7V.

4-8.1 Design example

Assume that the circuit of Fig. 4-23 is to track and hold a signal voltage in the range of 0 to 1V. The op-amp has an input bias current of 200 nA. The diodes have leakage of 1000 nA. It is desired to hold the maximum output voltage at 99 percent level for 3 seconds. For simplicity, assume that R_1 and R_2 are 10K, and that R_3 and R_4 are 1K.

Since the output must remain at 99 percent of maximum for 3 seconds, there must be a change no greater than 1 percent of the maximum for three seconds. Or, the output must be 0.99V for three seconds after the bridge is switched off ($1V \times 0.01 = 0.01V$; $1V - 0.01V = 0.99V$). Thus, the difference in output voltage is 0.01V (or 10 mV).

Using the equation of Fig. 4-23, the value of C_1 is:

$$C \approx \frac{3\ (1000 + 200) \times 10^{-9}}{0.01} \approx 360\ \mu\text{F}$$

Now assume that the value of C_1 is changed to 600 μF. All other conditions remain the same. Find the amount of time that the output will be held at the 99 percent level after the bridge is switched off, using the equation of Fig. 4-23.

$$t \approx \frac{600 \times 10^{-6}}{\left(\dfrac{1200 \times 10^{-9}}{0.01}\right)} \approx 5\ \text{seconds}$$

4-9. INTEGRATION AMPLIFIER (INTEGRATOR)

Figures 4-24 and 4-25 are working schematics of op-amps used as integration amplifiers (integrators). Integration of various signals (usually square waves) can be accomplished using these circuits. The output voltage from the amplifier is inversely proportional to the time constant of the feedback network and directly proportional to the integral of the input voltage.

The circuit of Fig. 4-24 is best suited to applications in which the integrator must be used over a wide range of frequencies. Keep in mind that the output amplitude will depend upon frequency, once the values are selected. The circuit in Fig. 4-25 is best suited where only one frequency is involved.

4-9.1 Design example

The value of R_1 is chosen on the basis of input bias current and voltage drop, and to produce realistic values for C_F and C_1. For simplicity, assume that R_1 is 10K.

The time constant of the $R_2 C_F$ combination must be substantially larger than the $R_1 C_F$ combination. For that reason, the values of R_{2A} and R_{2B} are approximately 10 times the value of R_1, or 100K, as indicated by the equations.

With the values of R_2 set, the value of C_2 is determined by the low frequency limit of the integrator (Freq. A). Typically, Freq. A is a fraction of 1 Hz. Assume a Freq. A of 0.016 Hz. Using these values, C_2 is approximately:

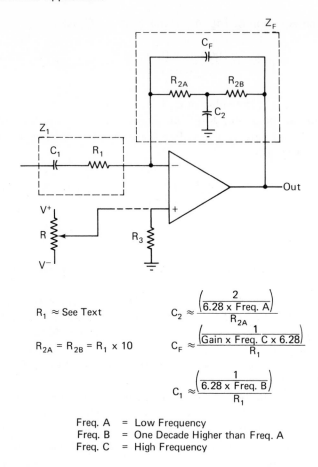

$R_1 \approx$ See Text

$R_{2A} = R_{2B} = R_1 \times 10$

$$C_2 \approx \dfrac{\left(\dfrac{2}{6.28 \times \text{Freq. A}}\right)}{R_{2A}}$$

$$C_F \approx \dfrac{\left(\dfrac{1}{\text{Gain} \times \text{Freq. C} \times 6.28}\right)}{R_1}$$

$$C_1 \approx \dfrac{\left(\dfrac{1}{6.28 \times \text{Freq. B}}\right)}{R_1}$$

Freq. A = Low Frequency
Freq. B = One Decade Higher than Freq. A
Freq. C = High Frequency

Fig. 4-24. Integration amplifier (integrator) for wide frequency range applications.

$$C_2 \approx \frac{\left(\dfrac{2}{6.28 \times 0.016}\right)}{100{,}000} \approx 200 \ \mu\text{F}$$

With the value of R_1 set, the value of C_F is determined by the high frequency limit of the integrator (Freq. C), and the gain desired at Freq. C. Assume that Freq. C is 5 kHz, and the desired gain is 10. Using these values, C_F is approximately:

$$C_F \approx \frac{\left(\dfrac{1}{10 \times 5000 \times 6.28}\right)}{10{,}000} \approx 300 \ \text{pF}$$

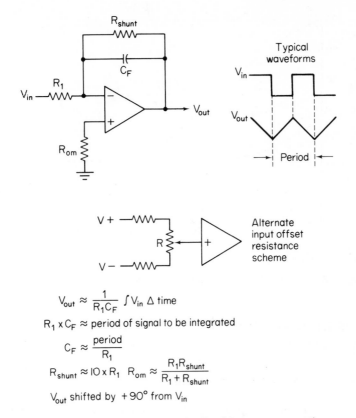

$$V_{out} \approx \frac{1}{R_1 C_F} \int V_{in} \, \Delta \text{ time}$$

$R_1 \times C_F \approx$ period of signal to be integrated

$$C_F \approx \frac{\text{period}}{R_1}$$

$R_{shunt} \approx 10 \times R_1 \quad R_{om} \approx \frac{R_1 R_{shunt}}{R_1 + R_{shunt}}$

V_{out} shifted by $+90°$ from V_{in}

Fig. 4-25. Integration amplifier (integrator) for fixed frequency operation.

With the value of R_1 set, the value of C_1 is determined by the intermediate frequency of the integrator (Freq. B). Note that Freq. B is approximately one decade above Freq. A. Assume a Freq. B of 0.16 Hz. Using these values, C_1 is approximately:

$$C_1 \approx \frac{\left(\dfrac{1}{6.28 \times 0.16}\right)}{10,000} \approx 100 \ \mu F$$

If the input offset voltage of the op-amp is low, the noninverting input can be connected to ground through a fixed resistance. As a first trial value, use the same resistance value as R_1. If greater precision is required, or if the op-amp offset is large, use the potentiometer network shown in Fig. 4-24. With this arrangement, potentiometer R is adjusted so that there is no offset voltage under no-signal conditions.

The voltage gain of the circuit shown in Fig. 4-24 is dependent upon frequency. Figure 4-26 shows a gain versus frequency response curve

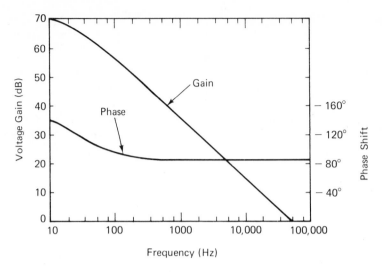

Fig. 4-26. Gain versus frequency response for integration amplifier.

for an op-amp (Motorola MC-1531) using the component values specified in this example.

4-9.2 Design example

In the circuit in Fig. 4-25, the value of R_1 is chosen on the basis of input bias current and voltage drop. Assume that the input bias current is 5000 nA and that the desired voltage drop across R_1 is not to exceed 180 mV (say, to provide 10 percent of a 1.8V input signal). A value of 33K for R_1 will produce a 165 mV drop. Assume an arbitrary value of 33K for R_1.

The value of the $R_1 C_F$ time constant must be *approximately* equal to the period of the signal to be integrated. The value of the $R_{shunt} C_F$ time constant must be *substantially larger* than the period (approximately 10 times longer). Thus, R_{shunt} is 10 times R_1. Note that R_{shunt} and C_F form an impedance that is frequency-sensitive (that is, the impedance is most noticed at low frequencies).

Assume that the circuit in Fig. 4-25 is to be used as an integrator for 1 kHz square waves. This requires a period of approximately 1 millisecond.

Any combination of C_F and R_1 can be used, provided the value of C_F times R_1 is approximately 0.001. Using the assumed value of 33K for R_1, the value of C_F is 0.03 μF.

The value of R_{shunt} must then be at least 330K. Note that the purpose of R_{shunt} is to provide direct-current feedback. This feedback is necessary so that an offset voltage cannot continuously charge C_F (which can result in amplifier limiting). If the offset voltage is small or can be minimized by means of R_{om}, it is possible to eliminate R_{shunt}.

Resistor R_{shunt} may have the effect of limiting gain at very low frequencies. However, above about 15 Hz, the effect of R_{shunt} is negligible (because of C_F in parallel).

If greater precision is required, particularly at low frequencies, the input offset resistance R_{om} can be replaced by the potentiometer network shown in Fig. 4-25. With this arrangement, potentiometer R is adjusted so that there is no offset voltage under no-signal conditions.

4-10. DIFFERENTIATION AMPLIFIER (DIFFERENTIATOR)

Figure 4-27 is the working schematic of an op-amp used as a differentiation amplifier (or differentiator). Differentiation of various signals (usually square waves, or sawtooth and sloping waves) can be accomplished using this circuit. The output voltage from the amplifier is inversely proportional to the feedback time constant, and is directly proportional to the time rate of change of the input voltage.

The value of the $R_F C_1$ time constant should be approximately equal to the period of the signal to be differentiated. In practical applications, the time constant must be chosen on a trial and error basis to obtain a reasonable output level.

The main problem in design of differentiating amplifiers is that the gain increases with frequency. Thus, differentiators are very susceptible to high-frequency noise. The classic remedy for this effect is to connect a small resistor (on the order of 50 ohms) in series with the input capacitor so that the high-frequency gain is decreased. The addition of the resistor results in a more realistic model of the differentiator because a resistance is always added in series with the input capacitance by the signal source impedance.

Conversely, in some applications a differentiator may be used to advantage to detect the presence of distortion or high-frequency noise in the signal. A differentiator can often detect hidden information that is not detected in the original signal.

Differentiation permits slight changes in input slope to produce very significant changes in output. An example of this feature is in determining the linearity of a sweep sawtooth waveform. Nonlinearity results from changes in slope of the waveform. Therefore, if nonlinearity is present, the differentiated waveform (amplifier output) indicates the points of

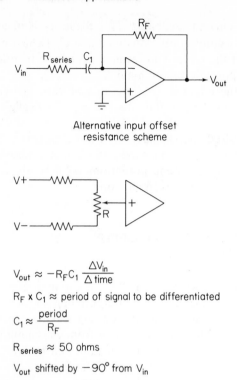

Alternative input offset
resistance scheme

Typical
waveforms

$$V_{out} \approx -R_F C_1 \frac{\Delta V_{in}}{\Delta \text{ time}}$$

$R_F \times C_1 \approx$ period of signal to be differentiated

$$C_1 \approx \frac{\text{period}}{R_F}$$

$R_{series} \approx 50$ ohms

V_{out} shifted by $-90°$ from V_{in}

Fig. 4-27. Differentiation amplifier (differentiator).

nonlinearity quite clearly. (However, it should be noted that repetitive waveforms with a rise and fall of differing slopes can show erroneous waveforms.)

4-11. LOW-FREQUENCY SINEWAVE GENERATOR

Figure 4-28 is the working schematic of an op-amp used as a low-frequency sinewave generator. This circuit is a parallel-T oscillator. Feedback to the noninverting input becomes positive at the frequency indicated in the equation. Positive feedback is applied at all times. The amount of positive feedback (set by the ratio of R_1 to R_2) is sufficient to cause the op-amp to oscillate. In combination with the feedback to the noninverting input, feedback to the inverting input can be used to stabilize the amplitude of oscillation.

The value of R_1 is approximately 10 times the value of R_2. The ratio of R_1 and R_2, as set by the adjustment of R_2, controls the amount of positive feedback. Thus, the setting of R_2 determines the stability of oscillation.

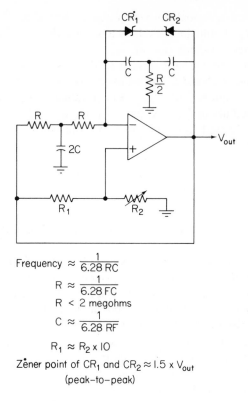

$$\text{Frequency} \approx \frac{1}{6.28\,RC}$$

$$R \approx \frac{1}{6.28\,FC}$$

$$R \;<\; 2 \text{ megohms}$$

$$C \approx \frac{1}{6.28\,RF}$$

$$R_1 \approx R_2 \times 10$$

Zener point of CR_1 and $CR_2 \approx 1.5 \times V_{out}$

(peak–to–peak)

Fig. 4-28. Low-frequency sinewave generator.

The amplitude of oscillation is set by the peak-to-peak output capability of the op-amp, and the values of Zener diodes CR_1 and CR_2. As shown by the equations, the Zener voltage should be approximately 1.5 times the desired peak-to-peak output voltage. The nonlinear resistance of the back-to-back Zener diodes is used to limit the output amplitude and maintain good linearity.

The frequency of oscillation is determined by the values of C and R. The upper-frequency limit is approximately equal to the bandwidth of the basic op-amp. That is, if the open-loop gain drops 3 dB at 100 kHz, the oscillator should provide full voltage output up to about 100 kHz.

4-11.1 Design example

Assume that the circuit in Fig. 4-28 is to provide 6V sinewave signals at 8 Hz.

Since R_2 is variable, the exact value is not critical. Assume a maximum value of 10K for R_2. With R_2 at 10K, R_1 is 100K.

With a required 6V peak-to-peak output, the values (Zener voltage) of CR_1 and CR_2 should be 9V. It is assumed that the basic op-amp is capable of 9V peak-to-peak output.

The values of R and C are related to the desired frequency of 8 Hz. Any combination of R and C can be used, provided that the combination works out to a frequency of 8 Hz. For practical design, the value of R should not exceed about 2M. Assume a value of 1M for simplicity. With R at 1M, and a desired frequency of 8 Hz, the value of C is:

$$C \approx \frac{1}{6.28 \times 8 \times 1^{+6}} \approx 0.02 \ \mu\mathrm{F}$$

4-12. PARALLEL-T FILTER (LOW FREQUENCY)

Figure 4-29 is the working schematic of an op-amp used as a *low-frequency filter*. The operating principle involved is similar to that of the parallel-T oscillator described in Sec. 4-11. However, the function is that of a narrow band filter (tuned peaking amplifier) described in Sec. 2-10 of Chapter 2. The circuit described in Sec. 2-10 uses an inductance as part of the resonant circuit. At very low frequencies, the high values of inductance (and capacitance) required make impractical a circuit similar to that described in Sec. 2-10. Thus, for low frequencies the parallel-T (or *twin*-T as it is sometimes called) filter is generally a better choice.

In the circuit of Fig. 4-29, gain is set by the open-loop gain of the op-amp. For this reason, the parallel-T filter is somewhat less stable than the tuned amplifier.

The frequency at which maximum gain occurs (or narrow band peak) is set by the values of R and C. Any combination of R and C can be used, provided that they work out to the desired frequency. However, the value of R should be selected on the basis of load resistance (or load impedance). The value of R and the load are, in effect, in parallel. If the value of R is many times (at least 10 times) that of the load, the net parallel resistance will be just slightly less than the load. Thus, the output current requirements for the op-amp are increased only slightly (for a given output voltage).

Figure 4-30 shows how the basic parallel-T filter circuit can be converted to a *low-frequency oscillator*. This is similar to the oscillator described in Sec. 4-11. Feedback to the noninverting input is obtained through the resistance network of R_3, R_4 and R_5. The amount of feedback, and thus the amplitude of the oscillator output, is controlled by

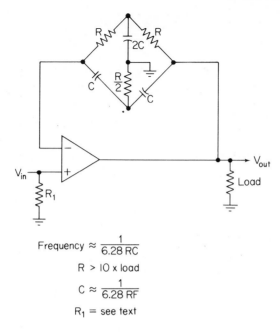

$$\text{Frequency} \approx \frac{1}{6.28\,RC}$$

$$R > 10 \times \text{load}$$

$$C \approx \frac{1}{6.28\,RF}$$

$$R_1 = \text{see text}$$

Fig. 4-29. Parallel-T filter (low frequency).

potentiometer R_4. Diodes CR_1 and CR_2 are used to limit the amount of feedback to the noninverting input. These diodes can be omitted with op-amps that can tolerate high input signals. The values for the resistance network shown in Fig. 4-30 are typical, and are not critical.

4-12.1 Design example

Assume that the circuit of Fig. 4-29 is to provide a peak frequency of 18 Hz, and that the load impedance is 20K.

The value of R_1 is chosen on the basis of input bias current and voltage drop, in most cases. However, with some applications, R_1 is chosen to provide a given input impedance, since filters must often work between two stages that require impedance match. Assume an arbitrary value of 10K for R_1.

With a load of 20K, the value of R is at least 200K. Use 200K for R. That is, both resistances shown as R on Fig. 4-29 should be 200K, each, and the single resistance shown as $R/2$ should be one-half 200K, or 100K.

With 200K for R, and a desired peak frequency of 18 Hz, the value of C is approximately 0.044 μF.

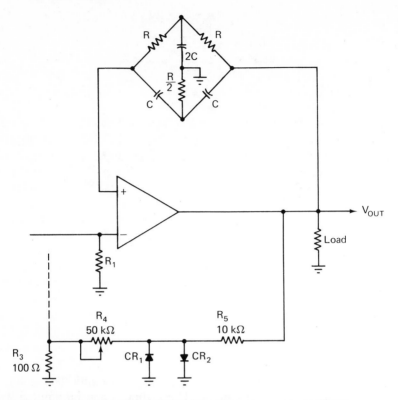

Fig. 4-30. Conversion of basic parallel-T filter to low-frequency oscillator.

Note that the circuit can be converted to an 18 Hz oscillator, using the resistance network shown in Fig. 4-30.

4-13. VOLTAGE REGULATORS

In this section, we shall discuss op-amps used as the amplifier or gain elements in voltage regulator systems. This application is not to be confused with IC packages that function as complete regulator circuits. Such packages are described in the author's *Manual For Integrated Circuit Users* (Reston Publishing Company, Reston, Va., 1973).

Regulators using op-amps as the gain elements usually show better regulation than packaged IC voltage regulators. This is primarily because of the higher available loop gain of the op-amp. Voltage drops of less than 0.01 percent over the entire load range are commonplace, and with care 0.001 percent is possible, using op-amps.

4-13.1 Basic regulator theory

Regulators can best be analyzed as a feedback system such as shown in Fig. 4-31. The output voltage and reference voltage in this theoretical system can be calculated using the equations shown. However, these equations do not take into account the output impedance of the op-amp, or the effects of the load.

The circuit in Fig. 4-32 is somewhat more practical. Here, the op-amp is followed by an emitter-follower stage, which provides necessary current gain (for a useful output current), as well as low output impedance. As shown by the equation, the output impedance of the regulator is approximately equal to the impedance seen at the base of Q_1, divided by the beta of Q_1. In turn, the base impedance is essentially the open-loop output impedance of the op-amp. Thus, the open-loop output impedance of the regulator is equal to the op-amp output impedance divided by the beta of Q_1.

When feedback is added, the regulator output impedance drops (by an amount proportional to open-loop gain), as shown by the closed-loop gain equation.

Using either circuit (Fig. 4-31 or 4-32), any variations in output voltage (due to changes in load), or variations in input voltage (changes in V_{ref}) cause a corresponding change in feedback voltage. In turn, this causes

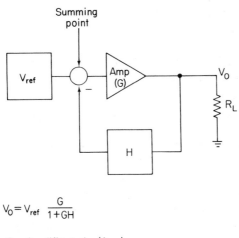

$$V_0 = V_{ref} \frac{G}{1+GH}$$

G = Amplifier gain (A_{vol})
H = Fraction of V_0 fed back to summing point

$$V_{ref} - V_0 = V_{ref} \frac{1}{1+G} \text{ for } H=1$$

Fig. 4-31. Theoretical feedback amplifier form of voltage regulator.

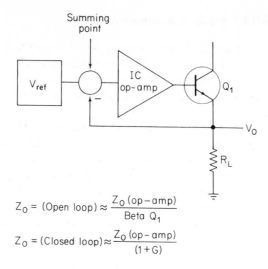

$$Z_O = (\text{Open loop}) \approx \frac{Z_O \,(\text{op}-\text{amp})}{\text{Beta } Q_1}$$

$$Z_O = (\text{Closed loop}) \approx \frac{Z_O \,(\text{op}-\text{amp})}{(1+G)}$$

Fig. 4-32. Theoretical feedback regulator with improved current capability.

the amplifier gain to change in a direction to oppose the initial change. For example, if V_{ref} is increased, V_o starts to increase as does feedback. The increase in negative feedback offsets the initial increase in V_{ref}, so the net effect is that V_o remains constant.

In a practical circuit, V_o is never truly constant. There is always some variation in output, which is expressed as a percentage of regulation. This is discussed in the next paragraph and, as is shown, the percentage of regulation decreases (regulation improves) when the closed-loop output impedance of the circuit decreases (and when op-amp gain increases).

4-13.2 Calculating percentage of regulation

The circuit in Fig. 4-33 can be used to calculate the theoretical percentage of regulation for any feedback amplifier. Figure 4-33 shows the regulator as a voltage source V', with a finite output impedance R_o. One equation for percentage of regulation involves the voltage source V' and the output voltage V_o. The alternate equation uses source V' and the term $I_L R_o$, which is the load current through the regulator output impedance. The $I_L R_o$ term is equivalent to the voltage drop across R_o, since $E = IR$. The term $I_L R_o$ also equals the difference between V' and V_o $(V' - V_o)$.

No matter which equation is used, the percentage of regulation is decreased (regulation is improved) when R_o is decreased. A decrease in R_o indicates a lower regulator system output impedance, a lower $I_l R_o$

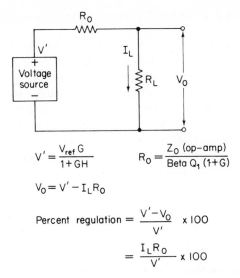

$$V' = \frac{V_{ref}\, G}{1+GH} \qquad R_0 = \frac{Z_0\ (op\text{-}amp)}{Beta\ Q_1\ (1+G)}$$

$$V_0 = V' - I_L R_0$$

$$\text{Percent regulation} = \frac{V'-V_0}{V'} \times 100$$

$$= \frac{I_L R_0}{V'} \times 100$$

Fig. 4-33. Theoretical regulator model.

term, a lower voltage drop across R_0, and a smaller difference between V' and V_0, all of which lower the percentage of regulation (improve regulation).

Using the theoretical circuit of Fig. 4-32, assume that the op-amp has an open-loop gain of 10,000, an output impedance of 800 ohms, that the emitter-follower has a beta of 30 and that V_{ref} is 18V.

Using the equation of Fig. 4-33, find the value of R_0

$$R_0 \approx \frac{800}{30 \times (1 + 10,000)} \approx 0.008 \text{ ohm}$$

Assuming a load current I_L of 300 mA, and using the $I_L R_0$ equation of Fig. 4-33, find the percentage of regulation:

$$\frac{0.300 \times 0.008}{18} \times 100 \approx 0.013 \text{ percent}$$

Using the $V' - V_0$ equation of Fig. 4-33, find the same percentage of regulation:

$$V_0 = V' - I_L R_0 = 18V - 0.0024 \approx 17.9976$$

$$\frac{V' - V_0}{V'} \approx \frac{18V - 17.9976}{18} \times 100 \approx 0.013 \text{ percent}$$

4-13.3 Effects of voltage offset on regulator action

Using the theoretical values of the previous example, the output voltage is supposed to follow the input or reference voltage within about 2.4 mV. This value could easily be masked by the input offset voltage of an op-amp, unless the input offset voltage is substantially less than 2.4 mV. Thus, unless some provisions are made to neutralize input offset voltage, the output voltage will always be removed from the input voltage by an amount equal to the offset. For example, assume that the uncorrected input offset voltage of the op-amp is 3 mV, and that the same reference (10V) as the previous example is used to find percentage of regulation:

$$\frac{18 - 17.997}{18} \times 100 \approx 0.016 \text{ percent}$$

Under these conditions, the percentage of regulation is limited by the input offset voltage, rather than the regulator system characteristics.

Another effect of input offset voltage is *output drift* with temperature and loading. Typical input offset voltage drift with temperature can be in the order of 5 μV/°C and higher. This not only appears as added shift between output voltage and reference voltage over a temperature range, but has variations in load as well.

For example, assume that the temperature increases 50°C for a given load change. If the drift is 20 μV/°C (not an unreasonable value for an op-amp), input offset voltage will change by a value of 1 mV (50×20 μV). Even if input offset voltage is neutralized to zero at the lower temperature, the output will be offset by 1 mV from the input, resulting in a 0.006 percent of regulation (assuming the values of the previous example).

In a practical op-amp regulator circuit, the output voltage will appear to drift after the load is applied or the load is changed. Then, after some time, the output voltage will settle to a constant value. The regulation figure using the equations of Fig. 4-33 will be of little value if the input offset voltage drift completely covers the short-term regulation. A typical op-amp requires about one to two minutes to settle after a change in temperature. This can easily interfere with good regulation for rapid load changes.

4-13.4 Effects of common mode rejection and power supply sensitivity on regulator action

If the op-amp receives its power directly from the regulator circuit input voltage as shown in Fig. 4-34, any variation in input voltage

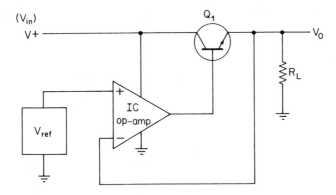

Fig. 4-34. Op-amp regulator powered by line voltage.

can produce some change in input offset voltage. The amount of input offset voltage change for a given power supply change is a function of power supply sensitivity. It is obvious that the percentage of regulation can be no better than the power supply sensitivity. For example, assuming the same 18V reference, an increase in input offset of 1 mV (due to power supply variation) produces about 0.006 percent of regulation change.

Another effect of power supply variation is that a common mode signal, or error voltage, is generated when V_{ref} remains constant (as it should), but the power supply voltages change. (The error is common mode since both power supplies $V_{CC} - V_{EE}$ or $V+$ and $V-$ change by the same amount.)

When V_{ref} is made the same as $V+$ and $V-$, the common mode error signal is equal to the difference between V_{ref} and the average of $V+$ and $V-$, or:

$$\text{common mode signal} = \left(\frac{V+ \; + \; V-}{2}\right) - V_{ref}$$

As a guideline, any op-amp used as the active element in a regulator should have a common mode rejection ratio of at least 90 dB, and preferably 100 dB.

4-13.5 Effects of voltage reference on regulator action

No matter what regulator circuit is used, output voltage stability over a period of time, and with temperature variations, depends for the most part on the quality of the voltage reference. An op-amp, even with very low temperature drift characteristics, cannot guarantee stable regulation. Thus considerable care must be used to select a good voltage reference.

Zener diode voltage reference. The most obvious solution is to use a Zener diode which has a low temperature coefficient by nature, or a Zener that has been compensated by adding forward diodes. Zener diodes have many problems. For example, low-voltage Zeners are difficult to use when the regulator must provide a high-voltage output. High-voltage compensated Zeners are expensive. Also, when the reference voltage is equal to the output voltage, current from the input voltage must be used to operate the Zener. This can cause some line ripple to feed through to the output.

If a Zener is to be used as the reference for an op-amp regulator, a circuit similar to that in Fig. 4-35 is generally satisfactory. The drop across series resistance R_S is equal to the input voltage, less the Zener voltage (which becomes V_{ref}). Note that Fig. 4-35 includes equations for trial values of the reference circuit. The following summarizes the use of these equations.

The Zener voltage should be about 0.7 times the lowest input voltage. Thus, if the input voltage varies between 10 and 15V, the Zener voltage

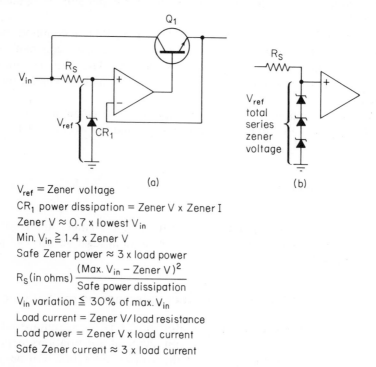

(a)   (b)

V_{ref} = Zener voltage

CR_1 power dissipation = Zener V × Zener I

Zener V ≈ 0.7 × lowest V_{in}

Min. $V_{in} \geqq 1.4$ × Zener V

Safe Zener power ≈ 3 × load power

R_S(in ohms) $\dfrac{(\text{Max. } V_{in} - \text{Zener V})^2}{\text{Safe power dissipation}}$

V_{in} variation $\leqq$ 30% of max. V_{in}

Load current = Zener V/load resistance

Load power = Zener V × load current

Safe Zener current ≈ 3 × load current

Fig. 4-35. Basic Zener voltage reference for op-amp regulator.

should be 7V (10V × 0.7). If the design problem is stated in reverse, the minimum input voltage must be 1.4 times the desired Zener voltage. Thus, if the required Zener (or V_{ref}) voltage is 10V, the minimum input voltage from the basic power supply must be 14V (10V × 1.4).

Because the Zener voltage must be about 30 percent below the input voltage (to ensure that the Zener will go into avalanche condition), there is a corresponding drop across the series transistor Q_1 when the Zener is used as V_{ref} (and the regulator output must equal V_{ref}). A large drop across Q_1 can create a power dissipation problem for Q_1.

Note that when a series of Zener diodes are used to achieve a given V_{ref} (Fig. 4-35b), the *total series Zener voltage* should be about 0.7 times the lowest input voltage.

A safe power dissipation rating for the Zener diode is 3 times the load power. Thus, with a 150 mW load, a Zener capable of dissipating 750 mW should be satisfactory. In the case of a regulator, the input resistance of the op-amp represents the load. Generally, op-amp resistance is quite high in relation to the voltage involved. Thus, the power dissipation rating of the Zener is low. Typically, a Zener (or series of Zeners) used as V_{ref} in an op-amp regulator requires a 0.5 to 1W power dissipation rating.

The value of R_S is found using the maximum input voltage, the desired Zener voltage, and the safe power dissipation. Using the same 10 to 15V input, a desired Zener voltage (V_{ref}) of 7V, and a safe power dissipation of 0.5W, the first trial value of R_S is:

$$\frac{(15 - 7)^2}{0.5} \approx 128 \text{ ohms}$$

In practical design, use the next highest value of 130 ohms.

An alternate method of using a Zener as V_{ref} is shown in Fig. 4-36. Here, an FET is used in place of R_S. Note that the FET gate and source terminals are shorted together. With this arrangement the FET will conduct when the drop across the FET exceeds the "pinch-off" voltage. Typically, the pinch-off voltage is on the order of 1 to 2V (or less). Thus, the input voltage need only be 1 to 2V higher than the desired Zener (or V_{ref}) voltage. This provides a minimum drop across the series transistor Q_1.

Variable voltage reference using an FET. One problem in using a Zener as V_{ref} is that the reference voltage is fixed. Thus, the voltage is dependent on available Zener values (either a single Zener or a series of Zeners). Likewise, the reference voltage cannot be adjusted. Often it is more desirable to have a regulator with a more flexible output. In a typical solid-state system, the power supply output (regulator output, in

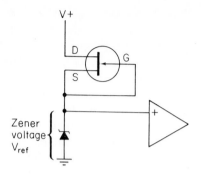

V + ≈ 1–2 V higher than Zenner voltage

Fig. 4-36. Zener voltage reference using FET in place of series resistance.

this case) should be capable of at least ±5 percent variation (preferably ±10 percent).

One inexpensive solution to this is shown in Fig. 4-37. Here, the voltage reference is maintained at a selected value by series resistors and an FET. The values shown are typical for input voltages up to about 20V, and possibly higher. The input voltage must be higher than the highest desired V_{ref} by about 1 to 2V (or less). This provides a minimum drop across the series transistor Q_1.

Potentiometer R_2 is adjusted for the desired V_{ref}. Any variation from this voltage is applied to the FET gate, and results in a corresponding variation in voltage drop across R_1. For example, if the input voltage goes up, V_{ref} goes up, as does the FET gate voltage. The FET draws more current, and the drop across R_1 increases to offset the initial change in V_{ref}.

The FET circuit of Fig. 4-37 cannot provide as stable a V_{ref} as the

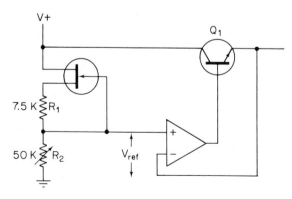

V + ≈ 1–2 V higher than V_{ref}

Fig. 4-37. FET variable voltage reference for op-amp regulator.

Zener circuits of Figs. 4-35 and 4-36. However, when adjustable V_{ref} is of greater importance, and cost must be kept at a minimum, the circuit in Fig. 4-37 will provide satisfactory results.

IC regulators as voltage reference. When V_{ref} must be adjustable, and temperature stable, the IC regulator packages provide the most satisfactory results. The basic circuit for using an IC regulator as V_{ref} in an op-amp regulator is shown in Fig. 4-38. Here, the V_{ref} is set by the ratio of R_1/R_2, and a constant that is a characteristic of the IC regulator package.

Using a typical example, assume that the constant is 3.5V, that R_2 is 10K, and that R_1 is variable from 10K to 40K. The V_{ref} is then variable from 7V to 17.5V: $(1 + 10K/10K) \times 3.5 = 7; (1 + 40K/10K) \times 3.5 = 17.5$.

The voltage range of IC regulators is typically zero up to 18V. However, there are IC regulators capable of handling up to about 37V.

A typical temperature drift specification for the IC regulators is 0.002%/°C. However, to realize the full value from the IC regulator, both resistors R_1 and R_2 must have temperature coefficients that match that of the regulator.

4-13.6 Operating at multiples of V_{ref}

When the regulator output must be considerably higher than the available V_{ref}, it is possible to operate at a multiple of V_{ref}, using

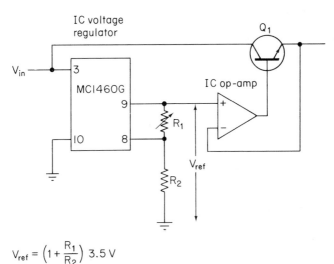

$$V_{ref} = \left(1 + \frac{R_1}{R_2}\right) 3.5 \text{ V}$$

3.5 V = constant of IC voltage regulator

Fig. 4-38. Basic op-amp voltage regulator using IC regulator as voltage reference.

the circuit of Fig. 4-39. As shown, the regulator output voltage V_o is set by the ratio of R_2/R_1, plus 1, times V_{ref}. Thus, if R_2 is four times the value of R_1, V_o is five times V_{ref}.

Of course, regulation for this circuit is worse than for a circuit where V_{ref} and V_o are equal, because of the loop attenuation introduced by R_1 and R_2. As shown by the equations, regulator output impedance Z_o, common mode rejection and power supply sensitivity are all degraded.

4-13.7 Floating op-amp regulators

Op-amp regulators are often used for stabilizing voltages considerably higher than the ratings of the op-amp. This control is possible because the entire circuit is not referenced to ground, but rather "floated" between ground and the supply voltage. Such a circuit is shown in Fig. 4-40. The circuit responds as a Zener "multiplier." The output V_o is a multiple of the reference voltage (across CR_3), and set by the ratio of R_1/R_2. Zener CR_1 ensures that the op-amp positive supply is greater than the required output swing. Zener CR_2 maintains a constant supply voltage for the op-amp.

The voltage across CR_2 is the sum of the positive and negative supply

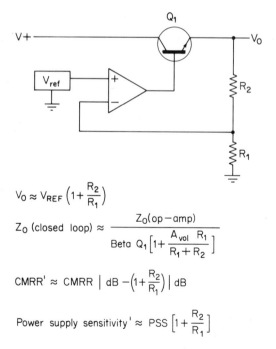

$$V_O \approx V_{REF}\left(1 + \frac{R_2}{R_1}\right)$$

$$Z_O \text{ (closed loop)} \approx \frac{Z_O(\text{op}-\text{amp})}{\text{Beta } Q_1\left[1 + \frac{A_{vol}}{R_1 + R_2}\right]}$$

$$\text{CMRR}' \approx \text{CMRR} \mid \text{dB} -\left(1 + \frac{R_2}{R_1}\right)\mid \text{dB}$$

$$\text{Power supply sensitivity}' \approx \text{PSS}\left[1 + \frac{R_2}{R_1}\right]$$

Fig. 4-39. Op-amp voltage regulator operating at multiples of V_{ref}.

voltages. With this arrangement, the voltage across the op-amp is tied to the output voltage by fixed constants. The voltage across CR_3 is equal to the output voltage, less the zero common mode level (which is the average of the $V+$ and $V-$ voltage levels).

Typically, the voltage across CR_1 is 10V, with the voltage across CR_2 equal to the sum of the supply voltages (if $V+$ and $V-$ are 15V, CR_2 must be 30V). Using these same values, and assuming a desired V_o of +75V, the remaining values can be found using the equations of Fig. 4-40 as follows:

With a V_o of 75V, and a V_{CR1} of 10V, the value of $V+$ is floated at

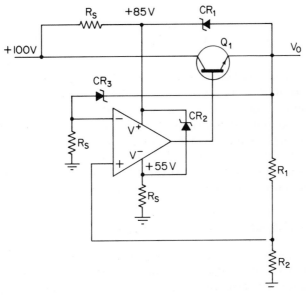

R_S = see text

$$V_0 \approx \frac{R_1 + R_2}{R_1} \times VCR_3$$

$$R_1 \approx \frac{VCR_3}{0.001\,A}$$

$$R_2 = \left(\frac{V_0}{0.001}\right) - R_1$$

Zero common mode level $\approx \dfrac{(V^+\text{ floating}) + (V^-\text{ floating})}{2}$

$VCR_1 \approx 10V$

$VCR_2 \approx V^+ + V^-$

$VCR_3 \approx V_0 -$ zero common mode level

V^+ floating $\approx VCR_1 + V_0$

V^- floating $\approx V^+$ floating $-(V^+ + V^-)$

$V^-\ V^+$ = supply voltages of IC

$V^-\ V^+$ floating = level in reference to ground

Fig. 4-40. Floating voltage regulator using op-amps.

+85V, with the $V-$ floating at +55V (30V, or the sum of the positive and negative supply voltages, below the $V+$ floating point).

With the $V+$ and $V-$ voltage levels (floating points) at +85V and +55V, respectively, the zero common mode level is 70V; $(85 + 55)/2 = 70$.

Thus, both the inverting and noninverting inputs to the op-amp are floated at 70V. With V_o at +75V, and the zero common mode level of +70V, the voltage across CR_3 is 5V $(75 - 70 = 5)$.

Assume an arbitrary 1 mA current through R_1 and R_2, or a total of 75K for $R_1 + R_2$ $(75/0.001 = 75,000)$. Using the equations, $R_1 = 5K$ $(5/0.001 = 5000)$. With $R_1 + R_2$ at 75K, and R_1 at 5K, $R_2 = 70K$.

The output V_o should then be:

$$V_o \approx \frac{R_1 + R_2}{R_1} \times V_{CR3} \approx \frac{5K + 70K}{5K} \times 5V \approx +75V$$

The value of R_S is selected on the basis of voltage drop and power dissipation, as described for Zener diode references (Sec. 4-13.5).

One limitation of the circuit in Fig. 4-40 is that regulation decreases (percentage of regulation increases) for increasing output voltage (all other factors being equal). Other limitations include the fact that there is a large voltage drop across R_2 and R_S in series with CR_3, as well as across the series transistor Q_1. However, the circuit can provide reasonably good regulation at voltages between about 100 and 250V.

4-13.8 Effects of ground loops

Thus far, we have discussed the effects of devices (op-amp, Zener, etc.) on overall regulator performance. In practical use, a well-designed regulator can show very poor performance if layout is poor, and little consideration is given to current paths. Typically, the regulator circuit elements will be mounted on the same board or card as the power supply. However, the regulator output will just as typically be delivered to a load at some remote location. In this sense, the term "remote" can mean another board or card only a few feet (or even a few inches) away from the regulator. In any event, the load current must pass through wires, which will produce some voltage drop.

A regulator diagram is given in Fig. 4-41 with critical load path wire resistances shown as R_{W1} and R_{W2}. These resistances are the most important simply because they carry the most current, and thus drop the most voltage.

Even if the output voltage V_o equals V_{ref}, and the op-amp gain is in-

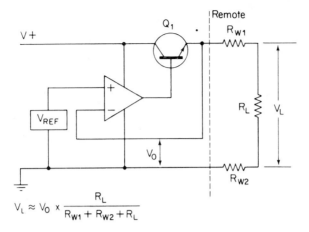

$$V_L \approx V_O \times \dfrac{R_L}{R_{W1} + R_{W2} + R_L}$$

Fig. 4-41. Basic ground loop model of op-amp regulator.

finite (neither of these conditions is practical), the actual voltage across the load V_L will still be a fraction of V_o. In effect:

$$V_L = V_o \times \frac{R_L}{R_{W1} + R_{W2} + R_L}$$

This may seem insignificant, but consider that No. 20 wire shows about 0.01 ohm of resistance per foot. This means about 1 mV per 100 mA of load current per foot. Assuming a 100 mA load (which is quite small for the typical power transistor used as series transistor Q_1), a V_{ref} of 10V, and a distance of 6 inches from the regulator card to the load card (12 inches total), the actual V_o will be offset by 1 mA from V_{ref}. This results in a *best possible* regulation of 0.01 percent.

Another problem arises if the regulator output is connected to the load by binding posts or plug-in terminals, rather than being soldered. Even solder joints can result in loss of a few millivolts if not properly made.

All of these conditions can be minimized (but not eliminated) by "remote sensing" as shown in Fig. 4-42. If the "sense" lines are the same length as the load lines, the voltage across the inverting and noninverting inputs to the op-amp are as shown by the equations in Fig. 4-42. In effect, both inputs are offset by the same amount. That is, the common mode voltage is increased by the drop across R_{W2}.

If the common mode rejection of the op-amp is good (at least 90 dB), the added common mode voltage will not significantly affect performance of the regulator.

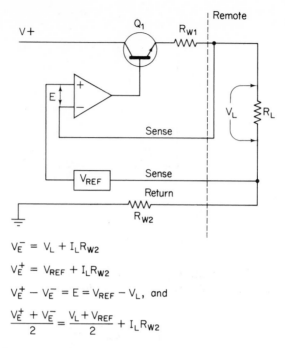

$$V_E^- = V_L + I_L R_{W2}$$

$$V_E^+ = V_{REF} + I_L R_{W2}$$

$$V_E^+ - V_E^- = E = V_{REF} - V_L, \text{ and}$$

$$\frac{V_E^+ + V_E^-}{2} = \frac{V_L + V_{REF}}{2} + I_L R_{W2}$$

Fig. 4-42. Basic op-amp voltage regulator with remote sensing.

The sense lines can be very small gauge wire since they carry very little current (compared to the load lines). The sense lines should be No. 20 wire or smaller. Resistance R_{W1} increases the open-loop output impedance of the regulator. However, the additional 0.01 to 0.02 ohm (for a load wire length of 1 foot) will not significantly affect performance.

4-13.9 A typical op-amp voltage regulator

A typical op-amp voltage regulator circuit is shown in Fig. 4-43, with a regulation curve shown in Fig. 4-44. Note that the output voltage drops less than 0.25 mV, with a load variation from 0 to 300 mA. Also note that the circuit components are identified as to type number or value. These values can be used as a "starting point" for design.

The regulator operates from a +20V source, and is adjusted for a +15V output by R_1. The drop across Q_1 is 5V and, with a 300 mA load, Q_1 must be capable of dissipating at least 1.5W (preferably more). Beta of Q_1 must be at least 20.

The reference voltage supplied by the IC regulator to the op-amp is equal to the ratio of R_1/R_2 times a constant of 3.5.

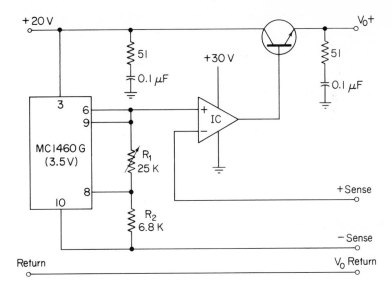

Fig. 4-43. Typical op-amp voltage regulator (Courtesy Motorola).

IC op-amp = MC 1539 G, or following characteristics
$A_{Vol} = 50,000$ (min)
$Z_0 = 4$ K
CMRR = 100 dB
Offset voltage = -4 mV max
$TC_{Vio} = 5$ $\mu V/°C$
Power supply sensitivity = 150 $\mu V/V$ (max)
$Q_1 = 2N4921$ with beta of 20 (min)

The characteristics of the op-amp are listed on Fig. 4-44. Note that the op-amp is operated from a single supply (+30V).

For best results, R_1, R_2 and Q_1 should be mounted close together so that they will all be at the same temperature. This will minimize the change in percentage of regulation with changes in temperature. R_1 and R_2 should also be matched in temperature characteristics to the IC regulator, and to each other.

In the final analysis, regulation will be no better than reference voltage stability, which, in turn, is dependent upon temperature characteristics.

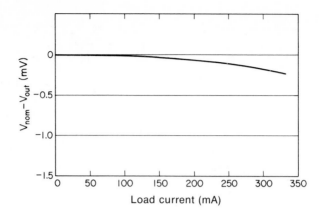

Fig. 4-44. Typical load regulation curve for op-amp regulator.

5. TESTING OPERATIONAL AMPLIFIERS

This chapter is devoted to test procedures for op-amps. These tests provide the user with a means of determining the actual or true characteristics of an op-amp. Finding true characteristics is more important than many users realize. There are several reasons for this.

When test information and characteristics are available for an op-amp, the values given are "typical." The datasheet values can vary from op-amp to op-amp, and with different operating conditions. There is no substitute for testing each op-amp under actual operating conditions (temperature, power supply, noise, etc.) of the intended use.

Keep in mind that a datasheet, or any other type of information, may not be available. More likely, the datasheet may not list all of the characteristics needed by the user. Even if all required characteristics are given, the procedures for finding the values may not be given, or are not clear.

Just as important, an op-amp can often be adapted to many uses other than the application intended by the manufacturer. Often the op-amp manufacturers are surprised at the uses to which their units are adapted. Therefore, they would have no reason for supplying test data (values or procedures) for such applications. The user must devise his own tests and find his own values.

Some op-amp manufacturers assume that all users will automatically know how to test for all characteristics. As a result, the manufacturers simply omit test data from their literature.

The following sections describe test procedures for op-amps. As a minimum, the following tests should be made on the op-amp, with power sources connected as in Chapter 1, operating in an open-loop circuit. The procedures can also be duplicated under closed-loop conditions if desired. The procedures will confirm (or deny) the op-amp characteristics found on the datasheet. The procedures can also be used to establish a set of characteristics for an op-amp when the datasheet is missing or inadequate.

It is assumed that the reader is thoroughly familiar with basic electronic test procedures. It is particularly important that the reader be familiar with the oscilloscope, voltmeter, signal generator and pulse generator.

5-1. FREQUENCY RESPONSE

The frequency response of an op-amp can be measured with a signal generator and a meter or oscilloscope. When a meter is used, the signal generator is tuned to various frequencies and the resultant op-amp output response is measured at each frequency. The results are then plotted in the form of a graph or *response curve,* such as shown in Fig. 5-1. The procedure is essentially the same when an oscilloscope is used to measure op-amp frequency response. However, an oscilloscope gives the added benefit of visual distortion analysis (as discussed in later sections of this chapter).

The basic frequency response measurement procedure (with either meter or oscilloscope) is to apply a *constant amplitude* signal while monitoring the op-amp output. The input signal is varied in frequency (but not amplitude) across the entire operating range of the op-amp. The voltage output at various frequencies across the range is plotted on a graph similar to that shown in Fig. 5-1. Generally, an oscilloscope is a better instrument for response measurement of an op-amp at higher frequencies. However, an electronic digital voltmeter can be used. Both open- and closed-loop frequency response should be measured with the same load.

1. Connect the equipment as shown in Fig. 5-1. Some manufacturers recommend that the load resistance be omitted. When the load resistance is used, it should be adjusted to the same value as the working load with which the op-amp is used. Also, some manufacturers recommend that the inverting input be connected directly to ground, rather than to a variable dc voltage. The purpose of the variable dc input is to set the output voltage offset to zero, or to some specific value. Set the generator, meter and/or oscilloscope operating controls as recommended in their respective instruction manuals.

2. If the variable input offset voltage is used, connect the op-amp output to the dc voltmeter, and adjust V_E until $V_{OUT(DC)}$ is zero, or at the voltage specified by the datasheet (typically in the order of ± 0.1V).

3. If a load resistance is used, adjust it to the same value as the intended op-amp load.

4. Connect the op-amp output to the ac voltmeter or to the oscilloscope. Initially, set the generator output frequency to the low end of the frequency range. Then set the generator output amplitude to the desired input level. For example, a typical op-amp may require 10 mV to produce full output voltage 5V.

5. In the absence of a realistic test input voltage, set the generator output to an arbitrary value. A simple method of finding a satisfactory input level is to monitor the circuit output (with the meter or oscilloscope) and increase the generator output at the op-amp center frequency (or at 1kHz) until the op-amp is overdriven. This point is indicated when further increases in generator output do not cause further increases in meter read-

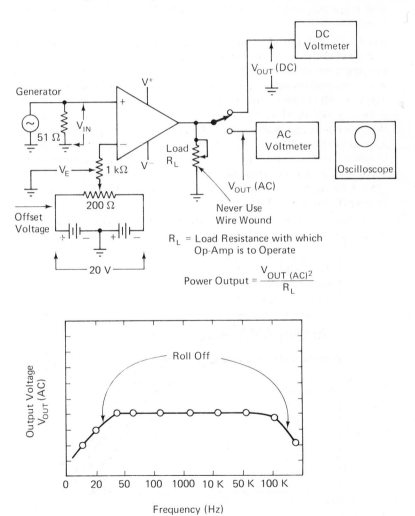

Fig. 5-1. Op-amp frequency response test connections and simplified response curve.

ing (or the output waveform peaks begin to flatten on the oscilloscope display). Set the generator output *just below* this point. Then return the meter or oscilloscope to monitor the generator voltage (at the op-amp input) and measure the voltage. Keep the generator at this voltage throughout the test.

6. Record the circuit output voltage on the graph. Without changing the generator output amplitude, increase the generator frequency by some fixed amount and record the new circuit output voltage. The amount of frequency increase between each measurement is an arbitrary matter. Use an increase of 10 Hz at the low end and high end (where rolloff occurs), and an increase of 100 Hz at the middle frequencies. If the op-amp frequency range is very broad, use 1-kHz increments at the middle frequencies.

7. Repeat the process, checking and recording the op-amp output voltage at each of the check points to obtain the frequency response curve. Depending upon op-amp design, the curve will resemble that of Fig. 5-1, with a flat portion across the middle frequencies, and rolloff at each end. Theoretically, there should be no rolloff at the low end, since an op-amp is direct-coupled. However, there is usually a capacitor at the output of the generator that combines with the op-amp input impedance or resistance to form a high-pass (low-cut) filter. This produces rolloff at the low-frequency end. For this reason, many frequency response graphs found on op-amp datasheets omit the low-frequency response. Figure 5-2 is a typical frequency response graph for an op-amp. Note that the lowest frequency shown is 1 kHz.

8. Note that generator output may vary with changes in frequency, a fact often overlooked in making a frequency response test of any circuit.

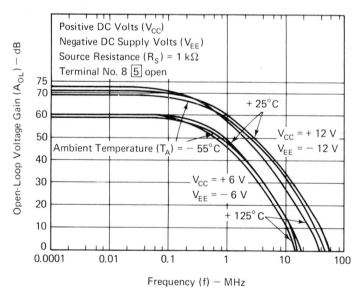

Fig. 5-2. Open-loop voltage gain versus frequency for CA3008 (Courtesy RCA).

Even precision laboratory generators can vary in output with changes in frequency, thus resulting in considerable error. Therefore, it is recommended that the generator output be monitored after each change in frequency (some generators have a built-in output meter). Then, if necessary, the generator output amplitude can be reset to the correct value. Within extremes, it is more important that the generator *output amplitude remain constant* rather than at some specific value when making a frequency response check.

5-2. VOLTAGE GAIN, BANDWIDTH AND PEAK-TO-PEAK OUTPUT VOLTAGE

Voltage gain measurement in an op-amp is made in the same way as frequency response. The ratio of output voltage to input voltage (at any given frequency, or across the entire frequency range) is the voltage gain. Since the input voltage (generator output) must be held constant for a frequency response test, a voltage gain curve should be identical to a frequency response curve.

Figure 5-3 shows a typical open-loop voltage gain curve for an op-amp. The voltage gain, shown as A_{OL} and given in dB, is found by:

$$20 \log \frac{V_{OUT}}{V_{IN}}.$$

(V_{OUT} and V_{IN} are shown in Fig. 5-1.)

Note the frequency at which the open-loop voltage gain drops 3 dB from the low-frequency value. This is the open-loop bandwidth (BW_{OL}). In the example of Fig. 5-3, the BW_{OL} is approximately $200 - 300$ kHz.

Remember that open-loop A_{OL} and BW_{OL} are characteristics of the op-amp, rather than the external circuit. Closed-loop gain is (or should be) dependent upon the ratio of feedback and input resistances, whereas closed-loop bandwidth is essentially dependent upon phase compensation techniques.

Closed-loop characteristics are generally lower than open-loop characteristics (voltage gain is lower, frequency response narrower, etc.). However, closed-loop characteristics are modifications of open-loop characteristics.

Also keep in mind that the op-amp has maximum input and output voltage limits, neither of which can be exceeded without possible damage to the op-amp and/or clipping of the waveform. If datasheet values are available, apply the maximum rated input and measure the actual output. Check the output for clipping at the maximum level. If clipping occurs, decrease the input until clipping just stops, and note the input voltage.

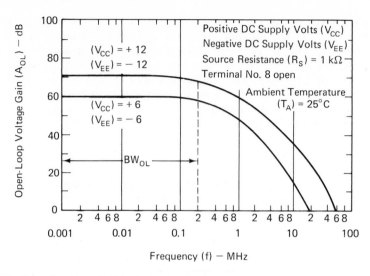

Fig. 5-3. Open-loop bandwidth (BW$_{OL}$) for CA3030 (Courtesy RCA).

Record these values as a basis for circuit design using the op-amp.

Note that maximum output voltage is dependent upon supply voltage and load resistance. This is shown in Fig. 5-4. For example, maximum peak-to-peak output voltage (shown as $V_o(P-P)$ in Fig. 5-4) is about 11.5V when the supply is ±12V, the load resistance is 10 kilohms, and the temperature is 25°C. If the supply and temperature remain constant, but the load resistance is increased to 15 kilohms, the $V_o(P-P)$ increases to about 13V.

5-3. CHANNEL SEPARATION OF IC OP-AMPS

Whenever two IC op-amps are put on the same monolithic chip, a certain amount of *crosstalk* will exist. That is, a signal applied to one amplifier will produce an output on the other amplifier, even though there is no signal applied to the second amplifier. This condition can be tested by applying a signal to one amplifier, and monitoring the output of the other amplifier, with the input to the second amplifier grounded.

The term *channel separation* is often used on dual-channel IC op-amps to show the relationship between the signal in the active channel (with input signal) and the signal in the inactive channel (with no input signal). Channel separation can be expressed in many ways. Figure 5-5 shows channel separation as a ratio between the output voltage of the active channel (in volts or V) and an induced input signal voltage to the inactive channel (in μV).

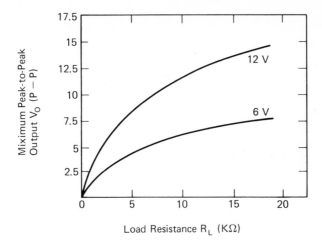

Fig. 5-4. Maximum peak-to-peak output voltage versus load resistance and supply voltage.

As shown in Fig. 5-5, channel separation is dependent upon closed-loop gain of both amplifiers and on frequency. In general, as frequency increases channel separation becomes poorer. To test for channel separation, connect both amplifier channels in the closed-loop configuration as shown in Fig. 5-5, apply a signal to one channel (channel #1), ground the input to the opposite channel (channel #2), and monitor the output of both channels. For convenience, adjust the input signal to channel #1 (and the ratio of feedback and source resistances R_A and R_B) so that the output of channel #1 is exactly 1V. Since the output of channel #2 will probably be in the microvolt range, with 1V at the output of channel #1, make the ratio of R_F to R_S at least 100 to 1, or 1000 to 1.

For example, assume that an op-amp has a channel separation curve similar to Fig. 5-5, that the input signal frequency is 300 kHz, that the output of channel #1 is set at 1V, and the ratio of R_F/R_S is 100-to-1. Under these conditions the output of channel #2 should be 20 mV, or 20,000 μV. With 20,000 μV at the output of channel #2, and a 100-to-1 ratio of R_F/R_S, this indicates an induced input signal e'_{in2} at channel #2 of 200 μV. Since the output of channel #1 is fixed at 1V, the channel separation ratio is 200 μV/V, which corresponds to the graph of Fig. 5-5. In some cases, channel separation is given in terms of dB rather than voltage.

5-4. POWER OUTPUT, GAIN AND BANDWIDTH

Most op-amps are not designed as power amplifiers. However, their power output, power gain and power bandwidth can be measured. The *power output* of an op-amp is found by noting the output

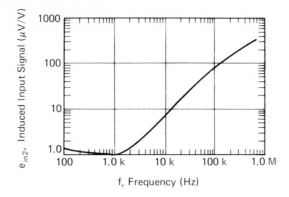

f, Frequency (Hz)

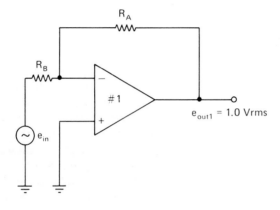

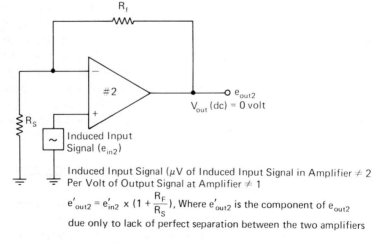

Induced Input Signal (μV of Induced Input Signal in Amplifier $\neq$ 2
Per Volt of Output Signal at Amplifier $\neq$ 1

$e'_{out2} = e'_{in2} \times (1 + \dfrac{R_F}{R_S})$, Where e'_{out2} is the component of e_{out2}

due only to lack of perfect separation between the two amplifiers

Fig. 5-5. Channel separation measurement (Courtesy Motorola).

voltage V_{OUT} across the load resistance R_L (Fig. 5-1.) at any frequency, or across the entire frequency range.

Power output is found by: $\dfrac{V_{OUT}^{2}}{R_L}$.

To find *power gain* of an op-amp, it is necessary to know both the input and output power. Input power is found in the same way as output power, except that the impedance at the input must be known (or calculated). This is not always practical in some op-amps, especially where impedance is dependent upon frequency or gain. The procedures for finding input impedance of an op-amp are discussed in later sections of this chapter. With input power known (or estimated) the power gain is found by:

$$\frac{\text{output power}}{\text{input power}}$$

Generally, power gain is not required by op-amp design specifications. Instead, an *input sensitivity* specification is often used. Input sensitivity specifications require a minimum power output with a given voltage input (such as 100-mW output with 10-mV RMS input).

Some power op-amp design specifications include a *power bandwidth* factor. Such specifications (generally limited to audio range op-amps) require that the op-amp deliver a given power output across a given frequency range. For example, an op-amp might produce full power output up to 20 kHz, even though the frequency response is flat up to 100 kHz. That is, voltage (without a load) will remain constant up to 100 kHz, whereas power output (across a normal load) will remain constant up to 20 kHz.

Figure 5-6 is the power bandwidth graph for an op-amp. The power bandwidth is given in terms of V_{OUT} (peak-to-peak) into a given load (2 kilohms), across the frequency range. In some op-amp specifications, power bandwidth is given as the frequency range in which a certain voltage swing can be accommodated at a given percentage of total harmonic distortion (such as $\pm 10V$ V_{OUT} with a 5 percent maximum total harmonic distortion). The subject of distortion is discussed in later sections of this chapter.

5-5. LOAD SENSITIVITY

Since an op-amp is generally not used as a power amplifier, load sensitivity is not critical. However, if it becomes necessary to measure load sensitivity use the following procedure.

An audio amplifier circuit of any type (including op-amp) is sensitive

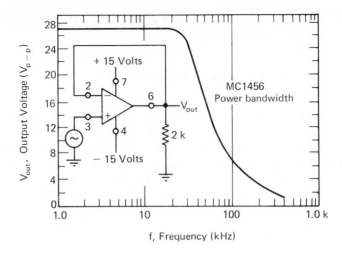

Fig. 5-6. Power bandwidth of op-amp (Courtesy Motorola).

to changes in load. This is especially true of power amplifiers. Any amplifier produces maximum power when the output impedance of the amplifier is the same as the load impedance. For example, if the load is twice the amplifier output impedance, the output power is reduced approximately 50 percent. If the load is 40 percent of the amplifier output impedance, the output power is reduced to approximately 25 percent. This is shown in Fig. 5-7 which is a typical load-sensitivity response curve.

The circuit for load sensitivity measurement is the same as for frequency response (Fig. 5-1), except that load resistance R_L is variable. Measure the power output at various load impedance/output impedance ratios. That is, set R_L to various resistance values, including a value for the amplifier output impedance, and note the voltage and/or power gain at each setting. Then repeat the test at various frequencies.

As indicated on the test diagram (Fig. 5-1), never use a wirewound load resistance. The reactance can result in considerable error. If a non-wirewound variable resistance of sufficient wattage is not available, use several fixed resistances (carbon or composition) arranged to produce the desired resistance values.

5-6. INPUT IMPEDANCE (SINGLE ENDED INPUT)

Dynamic input impedance of an op-amp can be found using either of the following procedures. Keep in mind that the closed-loop input impedance will differ from open-loop impedances. That is, closed-loop impedances depend upon feedback resistor values plus op-amp

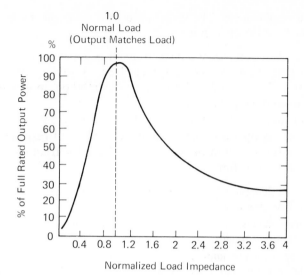

1.0
Normal Load
(Output Matches Load)

Fig. 5-7. Output power versus load impedance (showing effect of match and mismatch between output and load).

characteristics. Open-loop impedances depend solely on the op-amp circuit.

Ohmmeter-voltmeter method. Use the circuit of Fig. 5-8 to find the input impedance of an op-amp with an ohmmeter and voltmeter. The test conditions should be identical to those for frequency response, power output and so forth. That is, the same generator, operating load, meter or oscilloscope, and frequencies should be used.

1. Adjust the generator to the frequency (or frequencies) at which the op-amp will be operated.

2. Move switch S back and forth between positions A and B, while adjusting resistance R until the voltage reading is the same in both positions of the switch.

3. Disconnect resistor R from the circuit, and measure the dc resistance of

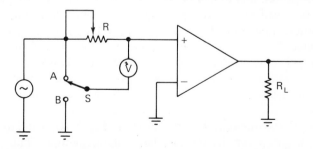

Fig. 5-8. Dynamic input impedance test connections (ohmmeter-voltmeter method).

R with an ohmmeter. The resistance of R is then equal to the dynamic impedance at the op-amp input.

Accuracy of the impedance measurement is dependent upon the accuracy with which the dc resistance is measured. A noninductive resistance must be used. The impedance found by this method applies only to the frequency used during the test.

Voltmeter method. Use the circuit shown in Fig. 5-9 to find the input impedance of an op-amp with a voltmeter and fixed resistance. Again, the test conditions should be identical to those for frequency response, power output and so forth.

1. Adjust the generator to the frequency (or frequencies) at which the op-amp will be operated.
2. Measure voltages V_A and V_B using a precision voltmeter. A digital voltmeter usually provides the most accurate results.
3. Calculate the input impedance Z_{in} using the equation in Fig. 5-9. For example, assume that V_A is 9V and V_B is 3V. Under these conditions:

$$Z_{in} = \frac{20,000}{\left(\frac{9}{3} - 1\right)} = 10,000$$

Accuracy of this impedance measurement is dependent upon accuracy of the voltmeter and the fixed resistances R_1 and R_2. Use resistors with a tolerance of ± 1 percent, or better.

5-7. OUTPUT IMPEDANCE

Dynamic output impedance of an op-amp can be found using either of the following procedures. Keep in mind that the closed-loop output impedance will differ from open-loop impedances, as is the case with input impedance.

Maximum power method. The load sensitivity test described in Sec. 5-5 can be reversed to find the dynamic output impedance of an op-amp. The connections and procedures (Fig. 5-1) are the same, except that the load resistance R_L is varied until *maximum output power* is found. Power is removed and R_L is disconnected from the circuit. The dc resistance of R_L (measured with an ohmmeter) is equal to the dynamic output impedance. The value applies only at the frequency of measurement. The test should be repeated across the entire frequency range.

Matched voltage method. Use the circuit in Fig. 5-10 to find the output impedance of an op-amp by the matched voltage method. The test con-

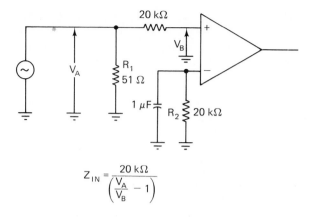

$$Z_{IN} = \frac{20 \text{ k}\Omega}{\left(\dfrac{V_A}{V_B} - 1\right)}$$

Fig. 5-9. Dynamic input impedance test connections (voltmeter method).

ditions should be identical to those for frequency response, power output and so forth.

1. Adjust the generator to the frequency (or frequencies) at which the op-amp will be operated.

2. With switch S_2 in position (c), adjust $V_E(DC)$ for a $V_{OUT}(DC)$ of 0 ± 0.1V.

3. With switch S_1 in position (a), and switch S_2 in position (d), record the voltage on the ac voltmeter as V_{OUT1} (rms).

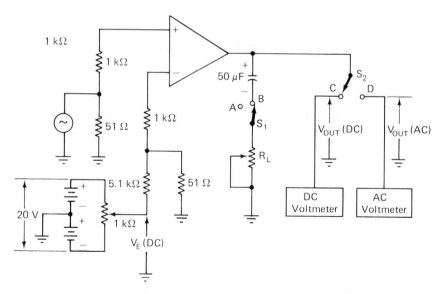

Fig. 5-10. Output impedance test connections (matched voltage method).

4. With switch S_1 in position (b), and switch S_2 in position (d), adjust R_L until the voltage on the ac voltemeter equals one-half of V_{OUT1} (rms).

5. Move switch S_1 back to position (a). Measure the dc resistance of R_L with an ohmmeter. The resistance of R_L is then equal to the dynamic impedance at the op-amp output.

Accuracy of this impedance measurement is dependent upon the accuracy with which the dc resistance is measured. A noninductive resistance must be used.

5-8. INPUT/OUTPUT CAPACITANCE

The capacitance between the input and output of an op-amp should be very small, typically a fraction of 1 pF. For this reason, input/output capacitance is generally not measured, and usually does not appear on op-amp datasheets. However, there are exceptions.

Figure 5-11 shows the input-to-output capacitance of a test circuit. Note that as supply voltage increases, input-to-output capacitance usually decreases. Also note that when measuring input-to-output capacitance the op-amp should be energized with normal supply voltages, and with both inputs shorted together, as shown by the schematic diagram in

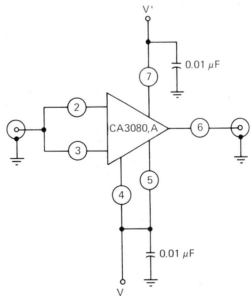

Fig. 5-11. Input-to-output capacitance test circuit (Courtesy RCA).

Fig. 5-11. A precision capacitance bridge is generally required for the measurement.

5-9. INPUT/OUTPUT LEAKAGE CURRENT

When the inputs and output of an op-amp are shorted together, and the op-amp is energized with supply voltages, there may be some leakage current from the shorted terminals through the op-amp circuits to ground, or to the supply voltage. Such leakage should be very small, typically less than 100 nA even at extreme temperatures, and generally less than 1 nA. For this reason, input/output leakage is generally not measured, and usually does not appear on op-amp datasheets.

Figure 5-12 is an exception and shows the input/output leakage of an op-amp (in graph form) as a function of temperature. Leakage to both ground (0V) and supply (36V) is shown. Note that the greatest amount of leakage occurs between the input/output terminals and the supply (36V) for a given temperature. It is also possible to measure leakage with the op-amp connected to the supply voltages in the normal manner ($+V$ and $-V$).

5-10. DISTORTION

Distortion requirements for op-amps are usually not critical. However, the four classic methods for measurement of distortion (sinewave analysis, squarewave analysis, harmonic analysis and intermodulation analysis) can be applied to any op-amp if desired. The following paragraphs summarize each of these methods.

5-10.1 Sinewave analysis

The procedure for checking op-amp distortion by means of sinewaves is to connect the equipment as shown in Fig. 5-1, and then monitor both the input and output waveforms with an oscilloscope. The primary concern is deviation of the op-amp output waveform from the input waveform. If there is no change (except in amplitude) there is no distortion. If there is a change in the waveform, the nature of the change will often reveal the cause of distortion.

In practice, analyzing sinewaves to pinpoint distortion is a difficult job,

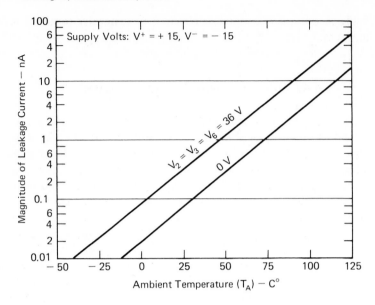

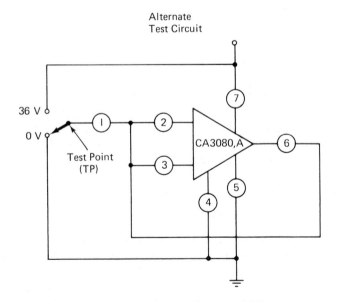

Fig. 5-12. Leakage current versus temperature (Courtesy RCA).

requiring considerable experience. Unless the distortion is severe it may pass unnoticed. Thus, if an oscilloscope is to be used alone (without intermodulation or harmonic distortion analyzers), square waves provide the best basis for distortion analysis.

5-10.2 Squarewave analysis

The procedure for checking distortion by means of square-waves is essentially the same as for sinewaves. Distortion analysis is more effective with squarewaves because of their high odd-harmonic content, and because it is easier to see a deviation from a straight line with sharp corners, than from a curving line.

As is the case of sinewave distortion testing, squarewaves are intro-duced into the op-amp input, with the output monitored on an oscillo-scope. (See Fig. 5-13.) The primary concern is deviation of the op-amp (shown as an IC) output waveform from the input waveform (which is also monitored on the oscilloscope). If the oscilloscope has a dual-trace feature, the input and output can be monitored simultaneously.

If there is a change in waveform, the nature of the change will often show the cause of distortion. For example, a comparison of the square-wave response at the output of an op-amp against "typical" patterns of Fig. 5-13 can show such faults as poor frequency response, overshoot, ringing, phase shift and emphasis or attenuated gain at certain frequencies.

The third, fifth, seventh and ninth harmonics of a clean squarewave are emphasized. If an amplifier passes a given frequency and produces a clean squarewave output, it is safe to assume that the frequency re-sponse is good up to *at least nine times* the squarewave frequency.

5-10.3 Harmonic analysis

No matter what amplifier circuit is used or how well the circuit is designed, there is always the possibility of odd or even har-monics being present with the fundamental. These harmonics combine with the fundamental and produce distortion, as is the case when any two signals are combined.

Commercial harmonic distortion meters operate on the *suppression* principle. As shown in Fig. 5-14, a sinewave is applied to the op-amp input, and the output is measured on the oscilloscope. The output is then applied through a filter that suppresses the fundamental frequency. Any output from the filter is then the result of harmonics. This output is also displayed on the oscilloscope where the signal can be checked for fre-quency to determine the harmonic content. For example, if the input is

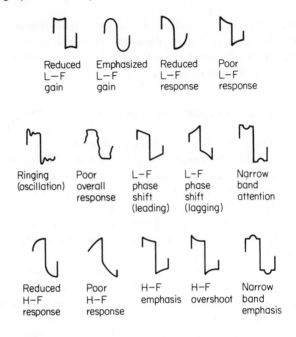

Reduced L–F gain   Emphasized L–F gain   Reduced L–F response   Poor L–F response

Ringing (oscillation)   Poor overall response   L–F phase shift (leading)   L–F phase shift (lagging)   Narrow band attention

Reduced H–F response   Poor H–F response   H–F emphasis   H–F overshoot   Narrow band emphasis

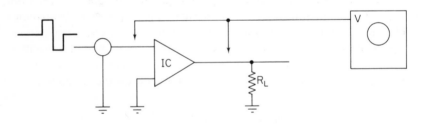

Fig. 5-13. Op-amp square wave distortion analysis.

1 kHz and the output after filtering is 3 kHz, third harmonic distortion is indicated. As shown in Fig. 5-14, third harmonic distortion results in flattening of the fundamental waveform peaks.

The percentage of harmonic distortion can also be determined by this method. For example, if the output without filter is 100 mV, and with filter is 3 mV, a 3 percent harmonic distortion is indicated.

The following steps describe the basic procedure for measurement of harmonic distortion. The exact procedure will depend upon the type of analyzer used. Always follow the analyzer manufacturer's operating instructions.

1. Connect the equipment as shown in Fig. 5-14.

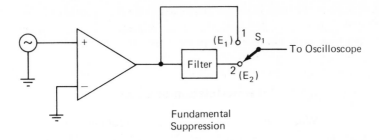

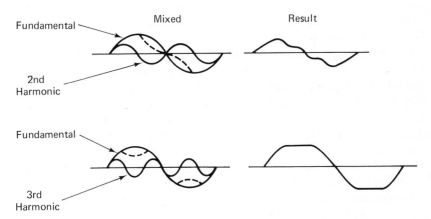

Fundamental
Suppression

$$\text{Percentage of Total Harmonic Distortion} = 100 \times \frac{E_2}{E_1}$$

Fig. 5-14. Harmonic distortion analysis.

2. Set the generator output frequency to the filter null frequency. In some commercial harmonic distortion analyzers, the filter is tunable so that the amplifier can be tested over a wide range of fundamental frequencies. In other analyzers, the filter is fixed frequency but can be detuned slightly to produce a sharp null.

3. Set the generator output amplitude to the maximum value specified in the op-amp datasheet. If specifications are not available, set switch S_1 to position 1, and increase the generator amplitude output until the waveform just starts to flatten, indicating that the amplifier is being overdriven. Then reduce the generator output until the waveform shows no distortion or flattening.

4. Measure the voltage with switch S_1 in position 1. Record this value as E_1.

5. Set switch S_1 to position 2. Adjust the filter for the deepest null indication on the oscilloscope. Record this value as E_2. (Note that some fundamental suppression filters are not tunable.)

6. Calculate the total harmonic distortion using the equation of Fig. 5-14.
7. If the filter is tunable, select another frequency, tune the generator to that frequency and repeat the procedures (steps 2 through 6).

5-10.4 Intermodulation analysis

When two signals of different frequency are mixed in an amplifier there is a possibility that the lower-frequency signal will amplitude-modulate the higher-frequency signal. This produces a form of distortion known as intermodulation distortion.

Commercial intermodulation distortion analyzers consist of a signal generator and high-pass filter as shown in Fig. 5-15. The signal generator portion of the analyzer produces a high-frequency signal (usually about 7 kHz for audio-range analyzers) that is modulated by a low-frequency signal (usually 60 Hz).

The mixed signals are applied to the amplifier input. The amplifier output is connected through a high-pass filter to the oscilloscope vertical channel. The high-pass filter removes the low-frequency (60 Hz) signal. Therefore, the only signal appearing on the oscilloscope vertical channel should be the high-frequency (7 kHz) signal. If any 60 Hz signal is present on the display, it is being passed through as modulation on the 7 kHz signal.

Figure 5-15 also shows an intermodulation distortion test circuit that can be set up when a commercial instrument is not available. The values and frequencies shown are suitable for a test of audio-range op-amps. The high-pass filter is designed to pass signals above approximately 200 Hz, and to reject signals below 200 Hz with a reduction of about 20 dB. If the op-amp must be tested at frequencies above the audio range, increase the generator and high-pass filter frequencies accordingly. The equation for determining filter values is shown in Fig. 5-15. The purpose of the 39 kilohm and 10 kilohm resistors is to set the 60-Hz modulating signal at four times that of the 7-kHz signal.

The following steps describe the basic procedure for measurement of intermodulation distortion. The exact procedure will depend on the type of analyzer used. Always follow the analyzer manufacturer's operating instructions.

1. Connect the equipment as shown in Fig. 5-15.
2. Set the generator frequencies to 7 kHz and 60 Hz, if the op-amp is to be tested in the audio range. Use other higher generator frequencies if necessary. The modulating frequency should be at least 100 times below that of the desired test frequency. For example, if the op-amp is tested

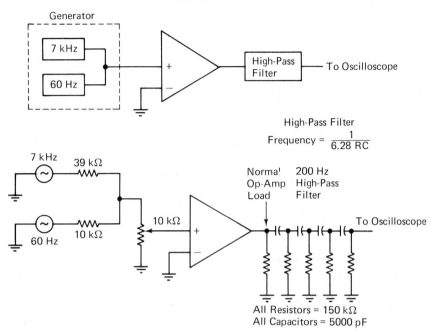

Basic Commercial Circuit

High-Pass Filter

$$\text{Frequency} = \frac{1}{6.28\ RC}$$

All Resistors = 150 kΩ
All Capacitors = 5000 pF

Modulation Envelope Display
on Oscilloscope

$$\text{Intermodulation Percentage} = 100 \times \frac{\text{Max} - \text{Min}}{\text{Max} + \text{Min}}$$

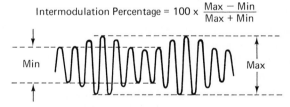

Fig. 5-15. Intermodulation distortion analysis.

for intermodulation distortion at 100 kHz, the modulating signal frequency should be no higher than 1 kHz (preferably below 1 kHz).

3. Set the generator output amplitude to the maximum value specified in the op-amp datasheet. If specifications are not available, increase the generator amplitude output until the waveform just starts to flatten, indicating that the amplifier is being overdriven. Then reduce the generator output until the waveform shows no distortion or flattening.

4. Measure the modulation envelope maximums and minimums to calculate the intermodulation distortion (using the equation of Fig. 5-15).

5. If desired, repeat the test using different frequency combinations.

5-11. NOISE TESTS

Op-amp noise tests can be very simple or very complex, depending upon requirements. There are two types of noise present in any op-amp. First, there is average background noise that remains constant over a period of time. Second, there is burst (or so called "popcorn") noise that is intermittent. We shall discuss both types of noise in the following paragraphs.

It must be noted that the noise measuring test equipment is highly specialized. For that reason, we shall not discuss the equipment in detail. Generally, the instructions supplied with noise test equipment are quite detailed, so we shall concentrate on understanding these instructions.

5-11.1 Background noise

The simplest form of background noise measurement is to monitor the output of an op-amp on a very sensitive voltmeter or oscilloscope, with the op-amp inputs either shorted to ground or terminated in resistances to ground. Any output voltage is background noise voltage (unless the input terminations are picking up stray noise around the op-amp). That is why noise tests should be conducted in a noise-free, shielded area. Usually, the op-amp is operated in the closed-loop configuration (typically with voltage gains of 100 to 1000), although some noise specifications require open-loop operation.

This simple type of noise test is generally not adequate. Instead, since background noise is usually frequency dependent, most background noise tests are made at a particular frequency, or band of frequencies. Such background noise is sometimes referred to as $1/f$ noise, indicating that the noise level is related to frequency. A tunable filter is connected between the op-amp and voltmeter. Usually, the filter is of the bandpass type. However, some noise tests use a low-pass filter, and show the noise at all frequencies below a given point. When noise is specified at one particular frequency, it is usually referred to as *spot noise*. On the other hand, *spectral noise* refers to the average noise over a band of frequencies, possibly around a given center frequency.

Figure 5-16 shows the spectral noise graph for a typical op-amp, together with the basic test connections. The op-amp is connected for a closed-loop voltage gain of 100. The voltmeter has a tunable range from 10 Hz to 100 kHz, with a bandwidth of 10 Hz. Note that the noise voltage drops as frequency increases. Noise voltage is also dependent upon bandwidth, input resistance and temperature.

The noise voltage, shown as e_n or equivalent noise, is given in nV/$\sqrt{\text{Hz}}$,

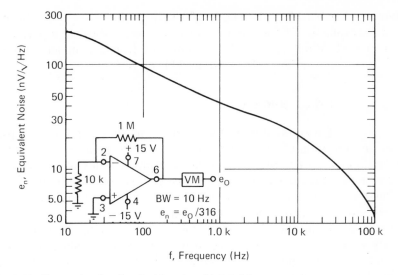

Fig. 5-16. Spectral noise density (Courtesy Motorola).

where $\sqrt{Hz}$ is the square root of the bandwidth frequency. Equivalent noise e_n is referred to the input, and is equal to the actual output voltage e_o divided by 316, as shown by the equation of Fig. 5-16. This equation is derived from the closed-loop gain of 100, and the 10 Hz bandwidth of the measuring voltmeter. The factor of 316 then refers the measured voltage back to the input and normalizes it to a 1 Hz bandwidth.

The graph of Fig. 5-16 can be used directly to find spot noise. Simply find the intersection of the graph curve at the frequency of interest. For example, at a frequency of 4 kHz the equivalent noise e_n is 30 nV. (The actual output voltage e_o is 9480 nV as measured on the voltmeter.)

The graph of Fig. 5-16 can also be used to find spectral noise, or noise over a given bandwidth at a given frequency. For bandwidths that are narrow compared to the frequency of interest, one may assume that the noise density is a constant. For example, to find the noise for a bandwidth of 100 Hz at a frequency of 10 kHz, find the intersection of the graph at 10 kHz (about 22 nV) and multiply the reading by the square root of 100 Hz, or 22 nV × 10 = 220 nV.

If the bandwidth is wide compared to the frequency of interest, the results are only approximate, at best. Wideband noise may be approximated by integrating or averaging the noise density over the entire band of interest. This is a difficult process, and can most easily be accomplished by assuming an average density over each octave, evaluating the total noise in each octave and then summing all the separate values.

The test connection of Fig. 5-16 can be duplicated for any op-amp. Of course, a tunable filter (10 Hz to 100 kHz, with a 10 Hz bandwidth) must be available. If background noise tests are critical to a particular op-amp application, best results are obtained from a commercial noise analyzer, such as a Quan-Tech Model 311, or equivalent.

5-11.2 Burst noise

Background noise measurement as shown in Fig. 5-16 provides an indication of the average noise power at the measurement frequency, but does not reveal burst noise characteristics of the op-amp. The metering circuits cannot respond fast enough to measure the effects of burst noise. Burst noise pulses are often very short in duration, although they can persist for several seconds.

The random rate at which the bursts occur can range from approximately several hundred per second to less than one per minute. The rates are not necessarily repetitive and predictable. Consequently, the nature of burst noise prevents its measurement by means of standard averaging techniques. Instead, a technique to detect individual bursts must be used, and the op-amp under test must be observed for a period on the order of 10 seconds to one minute.

RCA has developed test equipment to measure burst noise. A block diagram of the test equipment is shown in Fig. 5-17. Full details of the test equipment are available from RCA, and will not be repeated here. Instead, we shall summarize the equipment circuits to show how they relate to the tests.

The fixed high-gain amplifier portion of the circuit incorporates the op-amp as the first stage to amplify the microvolt-level burst to an easily detectable level. The test amplifier (post amplifier) is relatively free of burst noise.

The low-pass filter limits the test bandwidth to approximately 1 kHz. The reasons for selecting this particular bandwidth are described in later paragraphs.

The comparator produces a fast-rise, high-level, single-polarity output pulse whenever an input burst-noise pulse (of either polarity) exceeds a present (but adjustable) threshold level. The decade counter tallies the number of pulses from the comparator during the test period.

The latch circuit trips to the "latch" state when the count exceeds a preselected number. If tripped, the latch circuit energizes an indicator lamp. The timer determines the period during which the counter is enabled. The timer incorporates the capability to reset both the counter and the latch circuit at the beginning of each test period.

Three major characteristics of the noise burst affect op-amp applications: burst amplitude, duration and rate of occurrence. Of these, burst amplitude and rate of occurrence are of primary interest. Long duration bursts (of sufficient amplitude) seriously degrade the performance of dc amplifiers. However, suitable devices could be selected by the rejection of any unit that produced even one burst during some prescribed test period. Therefore, an absolute measurement of burst duration is not a prime necessity.

On the other hand, the rate of occurrence as measured by the burst-count in a given test period could conceivably be considered as a variable of prime importance. For example, a burst-rate of 100 per second is clearly objectionable in almost any low-level, low-frequency application, whereas the occurrence of only one low-amplitude burst in a one-minute

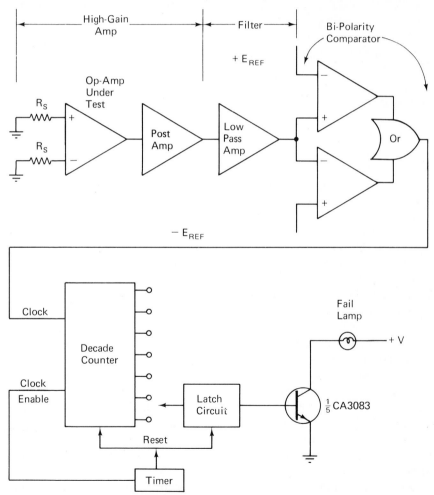

Fig. 5-17. Block diagram of RCA burst noise test set-up (Courtesy RCA).

period might be acceptable. The circuit in Fig. 5-17 detects total noise (1/f noise plus burst noise) bursts with amplitudes above a preset threshold level during a given test period and allows acceptance or rejection on the basis of the number of noise voltage excursions beyond the threshold level, in the selected test period.

Another factor to be considered is the bandwidth of the test system. Excessive bandwidth allows the normal noise of the terminating resistors and the op-amp to obscure burst-noise occurrences and does not realistically simulate the low-frequency applications in which burst-noise is particularly objectionable. On the other hand, a test circuit having excessively narrow bandwidth prevents detection of the shorter-duration bursts even if the amplitude is relatively high. Minimum duration bursts for an op-amp are generally in the order of 0.5 millisecond.

A suitable compromise is chosen in which the system rise time permits a burst of minimum duration to reach its full amplitude. Because the rise time and bandwidth of an amplifier are related by the equation *bandwidth* = $0.4/t_r$, the minimum bandwidth to detect a 0.5 millisecond burst is approximately:

$$\frac{0.4}{0.5 \times 10^{-3}} \approx 800 \text{ Hz}$$

Consequently, a 1 kHz bandwidth has been selected for the circuit in Fig. 5-17 as a reasonable one for a burst-noise test system.

Some of the other test conditions that affect the burst-noise performance of an op-amp include bias-level, source resistance R_S and ambient temperature.

The quiescent operating conditions in op-amps are normally set by the magnitude of the positive and negative supplies. Many of the newer op-amps, however, have bias terminals into which fixed currents can be injected to set the performance characteristics. The OTA units described in Chapter 3 are examples. For best low-frequency and burst-noise performance, such amplifiers should be operated at the lowest bias currents consistent with the gain-bandwidth requirements of the particular application.

In the test for burst noise, the source resistance R_S seen by the input terminals of the op-amp is a key test parameter. Burst noise causes effects that are equivalent to a spurious current source at the op-amp input. Therefore, burst-noise generates an equivalent input noise-voltage in proportion to the magnitude of the source resistance through which it flows.

To increase the sensitivity of the test system, it is desirable to use the highest source resistance consistent with the input offset current of the op-amp. For example, an op-amp that has 0.1 μA input offset current could realistically be tested with source-resistance on the order of 100 kilohms (10 mV input offset), whereas a 1 megohm source-resistance (100 mV input offset) could cause excessive offset in the output.

Burst noise generation in amplifiers is usually more pronounced at lower temperatures (particularly below 0°C). Consequently, consideration must be given to the temperature of the op-amp in relation to the temperature range under which the op-amp is expected to perform in a particular application.

Another test parameter of importance is the time duration of observation. Because the frequency of burst-noise occurrence is frequently less than one every few seconds, the minimum test period is from 15 to 30 seconds.

The burst-amplitude that will trip the counter can be no lower than the background noise peaks of typical burst-free op-amps. Otherwise, normal background noise will trip the fail indicator lamp. Therefore, the trip point is set just above the normal background noise. The average background noise can be determined by calculation or by direct measurement (at the output of the high gain amplifier-filter combination) using a storage oscilloscope or a "true RMS" voltmeter.

Selection of the acceptable number of burst counts in the test period is arbitrary, but dependent on the type of application intended for the op-amp. To be acceptable in some critical applications, the op-amp may not generate even a single burst-pulse in a relatively long period of time.

5-12. AVAILABLE FEEDBACK MEASUREMENT

Since op-amp characteristics are based on the use of feedback signals, it is often convenient to measure available feedback voltage at a given frequency with given operating conditions.

The basic feedback measurement connections are shown in Fig. 5-18. Although it is possible to measure available feedback voltage as shown in Fig. 5-18a, a more accurate measurement is made when the feedback lead is terminated in the normal operating impedance.

If an input resistance is used in the normal circuit, and this resistance is considerably lower than the op-amp input impedance, use the resistance value.

If in doubt, measure the input impedance of the circuit (Sec. 5-6), then terminate the feedback lead in that value to measure feedback voltage.

Keep in mind that when the feedback lead is disconnected (from the op-amp input) the output may change. Thus, the amount of feedback voltage will change. However, the test connections of Fig. 5-18 will permit measurement of available feedback voltage at a given frequency.

5-13. DIFFERENTIAL INPUT CURRENT

The circuit in Fig. 5-19 can be used to measure differential input current. This is not to be confused with input bias current that will be discussed in Sec. 5-14. Differential input current is the current that

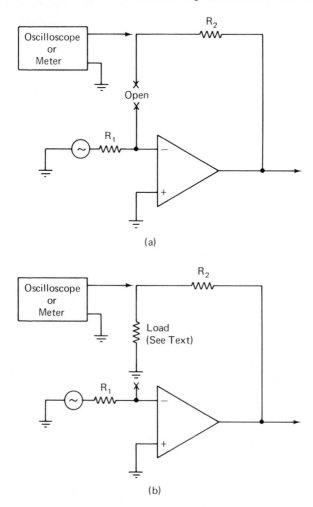

Fig. 5-18. Available feedback measurement.

flows between the differential inputs of an op-amp when a specific voltage is applied across the inputs.

As shown in the graph of Fig. 5-19, the differential input current increases with differential voltage and with temperature. In performing this test, make sure not to exceed the rated differential input of the op-amp.

5-14. INPUT BIAS CURRENT

Input bias current can be measured using the circuit in Fig. 5-20. Any resistance values can be used for R_1 and R_2, provided the value produces a measurable voltage drop. A value of 1000 ohms is realistic for both R_1 and R_2.

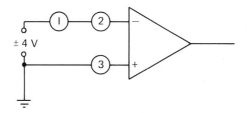

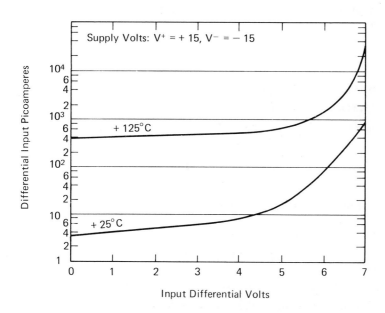

Fig. 5-19. Differential input current measurement (Courtesy RCA).

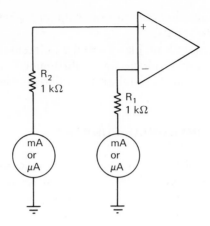

Fig. 5-20. Basic input bias current measurement.

Once the voltage drop is found, the input bias current can be calculated. For example, if the voltage drop is 3 mV across 1000 ohms, the input bias current is 3 μA.

In theory, the input bias current should be the same for both inputs. In practice, the bias currents should be almost equal for a well-designed op-amp. Any great difference in input bias is the result of unbalance in the input differential amplifier of the op-amp, and can seriously affect design.

A more precise measurement of input bias current can be made with the circuit shown in Fig. 5-21. Use the following procedure:

1. Adjust voltage V_E until V_{OUT} is zero, or less than 0.1V.
2. Measure and record V_{IN}.
3. Calculate the input bias current I_{IN} using the equation:

$$I_{IN} = \frac{V_{IN}}{100,000}$$

The input bias current can also be measured as described in Sec. 5-17.

5-15. INPUT OFFSET CURRENT

The circuit in Fig. 5-21 can also be used to measure input offset current. Use the following procedure:

1. Adjust voltage V_E until V_{OUT} is zero, or less than 0.1V.

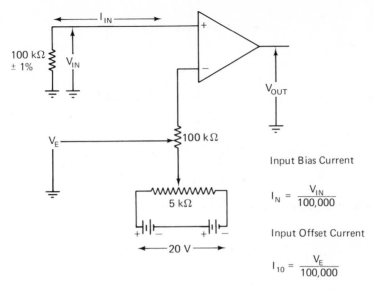

Input Bias Current

$$I_N = \frac{V_{IN}}{100,000}$$

Input Offset Current

$$I_{IO} = \frac{V_E}{100,000}$$

Fig. 5-21. Precision input bias current measurement.

2. Measure and record V_E.

3. Calculate the input offset current using the equation:

$$I_{IO} = \frac{V_E}{100,000}$$

Input offset current can also be measured as described in Sec. 5-17.

5-16. INPUT OFFSET VOLTAGE

The circuit in Fig. 5-22 can be used to measure input offset voltage. Simply measure V_{OUT} and divide by 100 (the op-amp gain). Input offset voltage can also be measured as described in Sec. 5-17.

5-17. INPUT OFFSET (VOLTAGE AND CURRENT) AND INPUT BIAS CURRENT

Input-offset voltage and current can be measured using the circuit in Fig. 5-23.

As shown, the output is alternately measured with R_3 shorted and with R_3 in the circuit. The two output voltages are recorded as E_1 (S_1 closed, R_3 shorted), and E_2 (S_1 open, R_3 in the circuit).

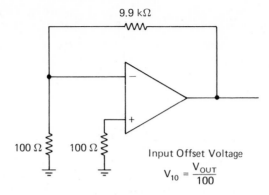

Fig. 5-22. Basic input offset voltage measurement.

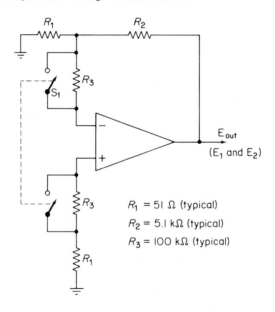

Fig. 5-23. Input offset voltage and current measurement.

With the two output voltages recorded, the input offset voltage and input offset current can be calculated using the equations of Fig. 5-23. For example, assume that $R_1 = 51$ ohms, $R_2 = 5100$ ohms, $R_3 = 100,000$, $E_1 = 83$ mV and $E_2 = 363$ mV (all typical values for an IC op-amp).

$$\text{Input offset voltage} = \quad 83 \text{ mV}/100 = 0.83 \text{ mV}$$
$$\text{Input offset current} = \frac{280 \text{ mV}}{100,000(1 + 100)} \approx 0.0277 \ \mu A$$

An alternate test circuit for input offset current and bias current measurement is shown in Fig. 5-24. Use the following procedures:

To measure inverting input bias current, set switch S_1 in closed position, and switch S_2 in open position. Measure output voltage V_{OUT} and convert this reading to inverting input bias current using the equation:

$$\frac{\text{Inverting}}{\text{bias current (in } \mu A)} = \frac{V_{OUT} \text{ (in volts)}}{10}$$

To measure non-inverting input bias current, set switch S_1 in open position, and switch S_2 in closed position. Measure output voltage V_{OUT} and convert this reading to noninverting input bias current using the equation:

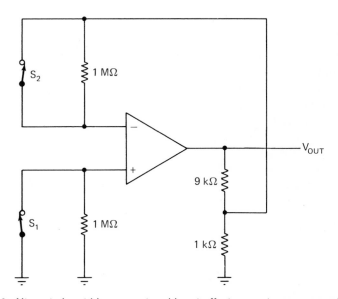

Fig. 5-24. Alternate input bias current and input offset current measurement.

$$\text{Noninverting bias current (in } \mu A) = \frac{-V_{OUT} \text{ (in volts)}}{10}$$

To measure input offset current, set switches S_1 and S_2 in open positions. Measure output voltage V_{OUT} and convert this reading to input offset current using the equation:

$$\text{Input offset current (in } \mu A) = \frac{V_{OUT} \text{ (in volts)}}{10}$$

5-18. POWER SUPPLY SENSITIVITY (INPUT OFFSET VOLTAGE SENSITIVITY)

Power supply sensitivity can be measured using the circuits in Figs. 5-22 and 5-23 (the same test circuits as for input offset voltage, Sec. 5-16 and 5-17).

The procedure is the same as for measurement of input offset voltage, except that one supply voltage is changed (in 1V steps) with the other supply voltage held constant. The amount of change in input offset voltage for a 1V change in one power supply is the power supply sensitivity (or *input offset voltage sensitivity,* as it is sometimes called).

For example, assume that the normal positive and negative supplies are 10V, and the input offset voltage is 7 mV. With the positive supply held constant, the negative supply is reduced to 9V. Under these conditions, assume that the input offset voltage is 5 mV. This means that the negative power supply sensitivity is 2 mV/V. With the negative power supply held constant at 10V, the positive supply is reduced to 9V. Now assume that the input offset voltage drops to 4 mV. This means the positive power supply sensitivity is 3 mV/V.

The test should be repeated over a wide range of power supply voltages (in 1V steps) if the op-amp is to be operated under conditions in which the power supply may vary by a large amount.

An alternate test circuit for power supply sensitivity is shown in Fig. 5-25. This test circuit can also be used to measure input offset voltage. Use the following procedures:

1. Adjust V_E for a V_{OUT} of $0 \pm 0.1V$.

2. Measure V_E and record input offset voltage in millivolts as:

$$\text{input offset voltage} = \frac{V_E}{1000}$$

3. With V_E set so that V_{OUT} is $0 \pm 0.1V$, increase V_{CC} by 1V (above the normal value) and record the new value of V_{OUT}.

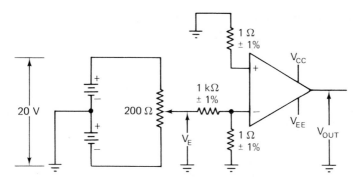

Fig. 5-25. Power supply sensitivity measurement.

4. Without changing V_E, decrease V_{CC} by 1V (below the normal value), and record the new value of V_{OUT}.

5. Divide the difference between V_{OUT} measured in steps 3 and 4 by the change in V_{CC} in steps 3 and 4, as follows:

$$\frac{V_{OUT}}{V_{CC}} = \frac{V_{OUT}(\text{step 3}) - V_{OUT}(\text{step 4})}{2 \text{ volts}}$$

For example, if V_{OUT} in step 3 is 70 mV, and V_{OUT} in step 4 is 30 mV, then the change in V_{OUT} is 40 mV. Under these conditions, the V_{OUT}/V_{CC} ratio is 40 mV/2V, or 20 mV/1V. This is generally shown on datasheets as 20 mV/V.

The value found in step 5 is the change in output voltage for a given power supply voltage change. Usually the desired value is the change in input voltage for changes in V_{CC}. That is, the measurement must be referred to the input. This can be done by dividing the change in V_{OUT} by the open-loop gain of the op-amp. For example, using the 20 mV/V figure, and an open-loop gain of 100, the input offset voltage sensitivity (or power supply sensitivity) is 0.2 mV/V.

6. Repeat the procedures of steps 3, 4 and 5 for the negative power supply V_{EE}. That is, hold V_{CC} at the normal rated value, set V_E so that V_{OUT} is $0 \pm 0.1V$, and then vary V_{EE} (above and below the normal value) by 1V.

5-19. COMMON MODE REJECTION

The common mode rejection of an op-amp can be measured using the circuit in Fig. 5-26. Before connecting the op-amp to the test circuit, measure the open-loop gain as described in Sec. 5-2, under specific conditions of frequency, input and so forth. This will establish op-amp gain when operating in the normal differential input mode. Use the same values of V_{CC} and V_{EE} for the common mode rejection test as for the open-loop gain test. Also, if the op-amp has a provision for input offset neutralization, adjust the input and output offset to zero before making the common mode rejection test.

1. With the op-amp connected as in the circuit in Fig. 5-26, adjust V_{BIAS} to zero (or remove the V_{BIAS} voltage). The V_{BIAS} is used only to measure common mode input voltage range as described in Sec. 5-20, and can be omitted if input voltage range is of no concern.

2. Increase the common mode voltage V_{IN} until a measurable V_{OUT} is obtained. Be careful not to exceed the maximum specified input common mode voltage swing. If no such value is specified, do not exceed the normal input voltage of the op-amp.

3. To simplify calculation, increase the input voltage until the output is 1 mV. Divide this value by the open-loop gain of the op-amp to find equivalent differential input signal. For example, with an open-loop gain of 100 (40 dB) and an output of 1 mV, the equivalent differential input signal is 0.00001 (10^{-5}). If the open-loop gain is 1000 (60 dB) and the output is 1 mV, the equivalent differential input signal is 10^{-6}.

4. Measure the input voltage V_{IN}. Divide V_{IN} by the equivalent differential input signal to find common mode rejection. For example, assume that V_{IN} is 0.3V with a V_{OUT} of 1 mV, and the open-loop gain is 100 (40 dB). The common mode rejection is:

$$\frac{0.001}{100} = 0.00001 \; ; \; 0.3/0.00001 = 30,000 \approx 90 \text{ dB}$$

The common mode rejection can also be calculated using the alternate:

$$\frac{0.001}{0.3} \approx 0.003; \; 0.003/100 = 0.00003 \approx 90 \text{ dB}$$

5-20. COMMON MODE INPUT VOLTAGE RANGE

The circuit shown in Fig. 5-26 can also be used to measure common mode input voltage range. First calculate the common mode rejection as described in Sec. 5-19. Then vary V_{BIAS} above and below 0V. Do not exceed the normal input voltage range of the op-amp.

The common mode input voltage range limits are those values of V_{BIAS} at which common mode rejection is 6 dB less than that calculated in Sec. 5-19. For example, if common mode rejection is 90 dB, the positive and negative values of V_{BIAS} that reduce common mode rejection to 84 dB represent the limits of common mode input voltage range. Generally, the positive and negative values are not the same. This is because the differential input of an op-amp is rarely balanced exactly.

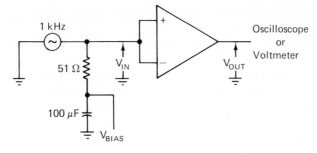

Fig. 5-26. Common mode rejection and common mode input voltage range measurement.

5-21. TRANSIENT RESPONSE

The circuit in Fig. 5-27 can be used to measure the transient response of an op-amp. Note that the unity gain configuration is used in Fig. 5-27 so that the output amplitude will be approximately equal to the input amplitude. The input signal must be low enough in amplitude to prevent saturation or limiting of the op-amp circuits. Also, the input signal (either squarewave or pulse), must have a *rise time that is less than the slew rate capability* of the op-amp. Otherwise, the squarewave or pulse input may be integrated into a sinewave. (Slew rate is discussed in Sec. 5-24.)

The values of C_L and R_L should match those of the load with which the op-amp is to be used. If load capacitance and load resistance are not established, use 100 pF for C_L and 2000 ohms for R_L.

Generally, the only op-amp transient response characteristics of any concern are *rise time* and *overshoot*. Both of these characteristics are measured by observing the output pulse from the op-amp on an oscillo-

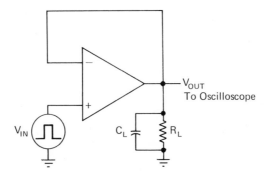

Fig. 5-27. Basic transient response measurement.

scope. Figure 5-28 shows a typical oscilloscope display of the pulse output from an op-amp.

Rise time is the time interval during which the amplitude of the output voltage changes from 10 percent to 90 percent of the rising portion of the pulse. In the waveform in Fig. 5-28, the rise time is approximately 0.3 µS.

Overshoot is a measure of the overshoot occurring above the 100 percent amplitude level of the output. Overshoot is generally expressed as a percentage of the amplitude of the rising portion of the pulse. In the waveform of Fig. 5-28, the pulse rises approximately 1 mV above the 100 percent amplitude level of 20 mV. To find the percentage, divide the overshoot (1 mV) by the maximum (20 mV) and multiply by 100, or (1/20) × 100 = 5 percent.

5-22. USING TRANSIENT RESPONSE TO FIND OP-AMP SYSTEM STABILITY

The response of an op-amp to a step (squarewave or pulse) input provides a measure of system stability. That is, the system stability of a compensated op-amp connected in a typical feedback configuration can be estimated using a simple technique based on measurement of the op-amp response to a squarewave or pulse input.

As discussed in Chapter 1, stability is one of the primary concerns in an op-amp system. Much time and effort is expended in a painstaking analysis (often undertaken with insufficient information about the characteristics of the op-amp) to confirm or ensure unconditional stability. This is usually followed by an extensive bench test involving time-consuming gain-bandwidth measurements. Frequently the op-amp data-sheet will specify the required compensation, but only for three or four

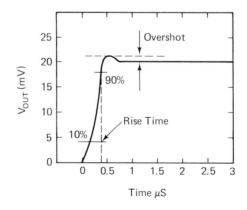

Fig. 5-28. Output voltage versus transient response time.

different gain configurations. In an effort to save time, the designer may use one of these worst case configurations, possibly resulting in the use of more restrictive compensation components than would actually be required.

The method described in this section was originally developed by Motorola. The method enables the designer to evaluate the design with a simple measurement, to make the indicated modifications (if any), repeat the measurement, and so forth until the desired performance is obtained.

5-22.1 Basis for the test method

Most op-amps have an uncompensated, open-loop frequency response that can be approximated by the form shown in Fig. 5-29, where the vertical scale is magnitude in dB, and the horizontal log scale is frequency (similar to those described in Chapter 1). When the open-loop op-amp is compensated for use as a feedback amplifier in a system with a closed-loop gain as low as unity, the slope of the compensated open-loop plot as it passes through zero dB must be less than 12 dB/octave or the closed loop system will be unstable. (If the open-loop slope is a constant 6 dB/octave, or can be so modified, as described in Chapter 1, the closed loop system is unconditionally stable, and the output waveform will have little or no overshoot, peaking, etc.)

For open-loop compensated op-amps that have a slope greater than 6 dB/octave, the closed-loop system is marginally stable, and will possibly have overshoot in response to step inputs, as well as peaking in frequency response. The most commonly used open-loop, compensated amplifier system that can be used in closed-loop systems with gains greater than or equal to unity, is the so-called two-pole system, which has a plot similar to that of Fig. 5-30. The test method discussed in this

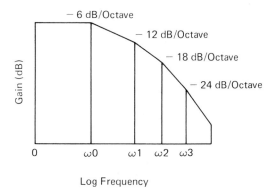

Fig. 5-29. Theoretical open-loop op-amp frequency response.

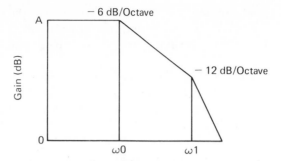

Fig. 5-30. Open-loop compensated op-amp response (two pole system).

section assumes that the op-amp will have an open-loop compensated response similar to that of Fig. 5-30 (where gain drops to zero somewhere in the 12 dB/octave region), and provides a means of determining stability over a given bandwidth when the op-amp is used in a closed-loop system. In addition to determining system bandwidth, the test method provides a close approximation of the amount of peaking.

5-22.2 Test conditions

To measure stability of a two-pole system using transient response, the op-amp must be connected in a test circuit that conforms as closely as possible to the actual operating condition. Figure 5-31 shows a typical test circuit. The op-amp is fully compensated using data-sheet values (or whatever compensation values are to be tested), and the feedback is set for a gain of 10. The input is from a squarewave or pulse generator. Output is connected to an oscilloscope.

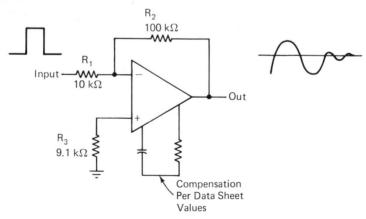

Fig. 5-31. Basic test circuit to measure system stability using transient response.

The input pulse may be simulated with use of a squarewave generator that has rise and fall times on the order of 100 nS (or some rise/fall times that exceed the slew rate capability of the op-amp). Thus, the output will not appear as a squarewave, but as an integrated wave similar to that shown in Fig. 5-32. Note that the output shown in Fig. 5-32 is the response to a single input pulse.

An alternative source would be a pulse generator capable of pulse widths on the order of 100 μS. Whatever pulse source is used, the generator must be terminated in a matching impedance (typically 50 ohms). Likewise, the generator should be capable of frequencies below 5 kHz.

During test, the pulse width and frequency controls of the generator must be adjusted to allow sufficient setting time for the op-amp to reach a steady-state (no-signal) condition before the next pulse appears. Also, the input amplitude must be small enough to prevent saturation of the op-amp (as is the case for transient response measurement described in Sec. 5-21).

$$Z = \frac{\text{Log } 10 \ \dfrac{V_{P1} - V_{P3}}{V_{P2}}}{\sqrt{\text{Log}^2 \ 10 \left(\dfrac{V_{P1} - V_{P3}}{V_{P2}}\right) + 1.8615}}$$

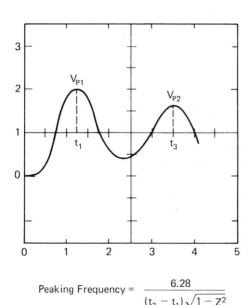

$$\text{Peaking Frequency} = \frac{6.28}{(t_3 - t_1)\sqrt{1 - Z^2}}$$

Fig. 5-32. Pulse response.

The key to accurate measurement using this method lies in the display of the output response on the oscilloscope. To achieve maximum accuracy, the response waveform should be displayed so that it occupies as much of the oscilloscope screen as possible. The oscilloscope grid or graticule is then scaled with the initial value of the waveform at zero, and the final value at 1, as shown in Fig. 5-32. The horizontal scale should be adjusted so that the three peaks are visible.

The values obtained from the oscilloscope display are substituted into the equations of Fig. 5-32 to determine peaking amplitude and frequency. The approximate peaking amplitude can be estimated from the values of Z. Figure 5-33 shows peaking as a function of Z. The table included on Fig. 5-33 lists approximate peaking in dB for selected values of Z.

5-22.3 Test example

The following is an example of how transient response can be used to measure system stability. Both peaking frequency and peaking amplitude are calculated, using the op-amp in a test configuration equivalent to that of the intended application. If the bandwidth indicated by these calculations provides the desired frequency range, the op-amp can be considered as stable over that bandwidth, at least.

Assume that the op-amp is connected in the test circuit of Fig. 5-31, and response is similar to that of Fig. 5-32. That is, the peak voltage (V_P) and time (t) values are: $V_{P1} = 1.867$, $V_{P2} = 0.366$, $V_{P3} = 1.534$, $t_1 = 1.16\ \mu\text{S}$, $t_3 = 3.48\ \mu\text{S}$.

Substituting these values into the equations of Fig. 5-32 yields:

$$Z = \frac{\left|\log_{10} \dfrac{1.867 - 1.534}{0.366}\right|}{\sqrt{\log^2_{10}\left(\dfrac{1.867 - 1.534}{0.366}\right) + 1.8615}} = 0.029$$

$$\text{Peaking frequency} = \frac{6.28}{3.48 - 1.16\ \sqrt{1 - Z^2}} \approx 431\ \text{kHz}$$

As shown in Fig. 5-33, a Z of 0.029 produces a peaking amplitude of about 24 dB. Thus, the frequency response of the op-amp should be similar to that of Fig. 5-34. That is, the response should be relatively flat at a voltage gain of 10 (20 dB), but there will be a peak of 24 dB at about 431 kHz. The frequency response shown in Fig. 5-34 is typical for an op-amp operated where the closed-loop gain crosses zero at a frequency in the 12 dB/octave region of open-loop response.

If the peaking frequency and/or amplitude are not within tolerance, change the op-amp compensation values and repeat the test as necessary until the peaking is within tolerance. Generally, peaking amplitude should be at a minimum, whereas peaking frequency should be at a maximum. However, bandwidth and peaking are dependent upon the intended application. Once the final compensation has been selected, and the

Table A

Z	P $\approx$ dB Peaking
0.01	33.9803
0.02	27.9609
0.03	24.4412
0.04	21.9454
0.05	20.0111
0.06	18.4323
0.07	17.099
0.08	15.9457
0.09	14.9301
0.1	14.0232
0.15	10.5565
0.2	8.13619
0.3	4.84662
0.4	2.69544
0.5	1.2494
0.6	0.35458
0.7	0.00173

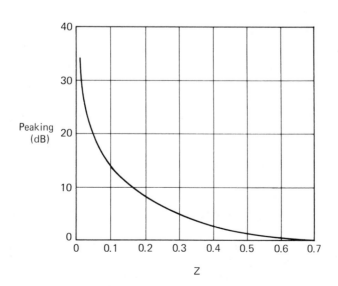

Fig. 5-33. Peaking (dB) as a function of Z (Courtesy Motorola).

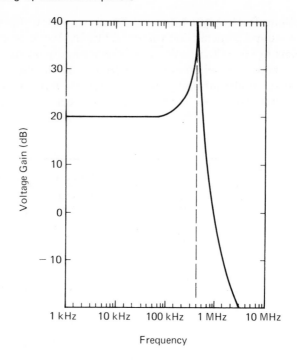

Fig. 5-34. Frequency response with peaking.

transient response is well within tolerance, it is advisable to test the system frequency response as described in Sec. 5-1 and 5-2.

5-23. PHASE SHIFT

Because an op-amp uses the principle of feeding back output signals to the input, the phase shift between input and output is quite critical. All of the op-amp phase compensation schemes are based on the feedback principle. Under ideal open-loop conditions, the output should be 180° out-of-phase with the (inverting) negative input, and in-phase with the positive (noninverting) input. As discussed in Chapter 1, the phase shift in a practical op-amp does not remain constant with changes in frequency. Generally, phase shift increases with frequency. In any event, it is often necessary to measure the phase shift of an op-amp to determine many factors (frequency response, best phase compensation scheme, etc.).

The following paragraphs describe two procedures for the measurement of phase shift between input and output of an op-amp. The procedures apply to both inputs (inverting and noninverting, negative and positive). The procedure can be used for any op-amp, provided the signals are of a frequency that can be measured on an oscilloscope.

The oscilloscope is the ideal tool for phase measurement. The most convenient method requires a dual-trace oscilloscope, or an electronic switching unit, to produce a dual trace. If neither of these is available, it is still possible to provide accurate phase measurements up to about 100 kHz using the single-trace of X-Y method.

5-23.1 Dual-trace phase measurement

The dual-trace method of phase measurement provides a high degree of accuracy at all frequencies, but is especially useful at frequencies above 100 kHz where X-Y phase measurements may provide inaccurate results (owing to inherent internal phase shift of the oscilloscope).

The dual-trace method also has the advantage of measuring phase differences between signals of different amplitudes and waveshapes, as is usually the case with input and output signals of an op-amp. The dual-trace method can be applied directly to those oscilloscopes having a built-in dual-trace feature or to a conventional single-trace oscilloscope using an electronic switch (known as a "chopper"). Either way, the procedure is one of displaying both input and output signals on the oscilloscope screen simultaneously, measuring the distance (in screen scale divisions) between related points on the two traces, then converting this distance into phase.

The test connections for dual-trace phase measurement are shown in Fig. 5-35. For the most accurate results, the cables connecting input and output signals should be of the same length and characteristics. At higher frequencies, a difference in cable length or characteristic could introduce a phase shift.

The oscilloscope controls are adjusted until one cycle of the input signal occupies exactly nine divisions (9 cm horizontally) of the screen. Then the phase factor of the input signal is found. For example, if 9 cm represents one complete cycle or 360°, 1 cm represents 40° (360/9 = 40).

With the phase factor established, the horizontal distance between corresponding points on the two waveforms (input and output signals) is measured. The measured distance is multiplied by the phase factor of 40°/cm to find the exact amount of phase difference. For example, assume a horizontal difference of 0.6 cm with a phase factor of 40° as shown in Fig. 5-35. Multiply the horizontal difference (0.6 cm) by the phase factor (40°/cm) to find phase shift (24°) between input and output signals.

Note that the test connections shown in Fig. 5-35 show the input signal applied to the inverting input. This will produce a 180° phase shift between input and output of a normal op-amp. Thus, if the test connections in Fig. 5-35 show a phase shift of, say, 160°, the phase shift is only 20° from normal.

More accurate phase measurements can be made if the oscilloscope is provided with a sweep magnification control by which the sweep rate can be increased by some fixed amount (5X, 10X, etc.) and only a portion of one cycle can be displayed. In this case, the phase factor and the approximate phase difference is found as described. Without changing any other controls, the sweep rate is increased (by the sweep magnifica-

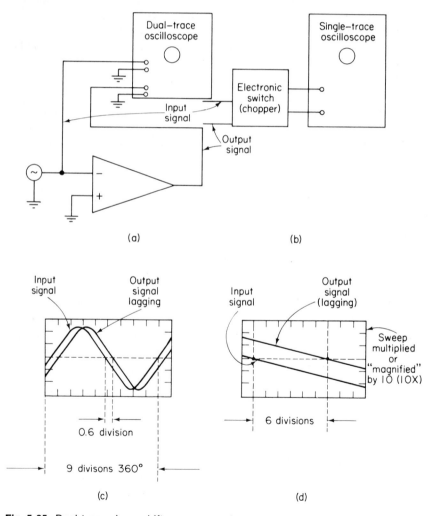

Fig. 5-35. Dual-trace phase shift measurements.

tion control or the sweep rate control) and a new horizontal distance measurement is made, as shown in Fig. 5-35d.

For example, if the sweep rate is increased 10 times, the adjusted phase factor is $40°/10 = 4°/cm$. Figure 5-35d shows the same signal as used in Fig. 5-35c, but with the sweep rate set to 10X. With a horizontal difference of 6 cm, the phase difference is $6 \times 4° = 24°$.

5-23.2 Single-trace (X-Y) phase measurement

The single-trace (or X-Y) phase measurement method can be used to measure the phase difference between input and output of an op-amp, at frequencies up to about 100 kHz. Above this frequency, the inherent phase shift (or difference between the horizontal and vertical systems of the oscilloscope) makes accurate phase measurements difficult.

In the X-Y method, one of the signals (usually the input) provides horizontal deflection (X), and the other signal provides the vertical deflection (Y). The phase angle between the two signals can be determined from the resulting pattern. The test connections for single-trace phase measurement are shown in Fig. 5-36.

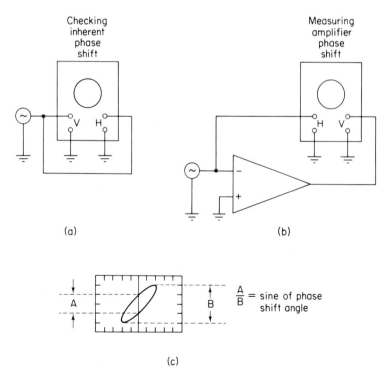

Checking
inherent
phase
shift

Measuring
amplifier
phase
shift

(a)

(b)

$$\frac{A}{B} = \text{sine of phase shift angle}$$

(c)

Fig. 5-36. Single-trace (X-Y) phase shift measurement.

Figure 5-36a shows the test connection necessary to find the inherent phase shift (if any) between the horizontal and vertical deflection systems of the oscilloscope. Inherent phase shift (if any) should be checked and recorded. If there is excessive phase shift (in relation to the signals to be measured), the oscilloscope should not be used. A possible exception exists when the signals to be measured are of sufficient amplitude to be applied directly to the oscilloscope deflection plates, bypassing the horizontal and vertical amplifiers.

The oscilloscope controls are adjusted until the pattern is centered on the screen as shown in Fig. 5-36c. With the op-amp output connected to the vertical input, it is usually necessary to reduce vertical channel gain (to compensate for the increased gain through the amplifier). With the display centered in relation to the vertical line, distances A and B are measured, as shown in Fig. 5-36c.

Distance A is the vertical measurement between two points where the traces cross the vertical centerline. Distance B is the maximum vertical height of the display. Divide A by B to find the *sine of the phase angle* between the two signals. This same procedure can be used to find inherent phase shift (Fig. 5-36a) or phase angle (5-36b).

Figure 5-37 shows the displays produced between 0° and 360°. Notice that above a phase shift of 180°, the resultant display will be the same as at some lower frequency. Therefore, it may be difficult to tell whether the signal is leading or lagging. One way to find correct polarity (leading or lagging) is to introduce a small known phase shift into one of the signals. Then the proper angle may be found by noting the direction in which the pattern changes.

As shown in Fig. 5-37, if the display appears as a diagonal straight line, the two signals are either in-phase (tilted from upper right to lower left) of 180° out-of-phase (tilted from upper left to lower right). These are the ideal conditions for an op-amp. If the display is a perfect circle, the signals are 90° out-of-phase.

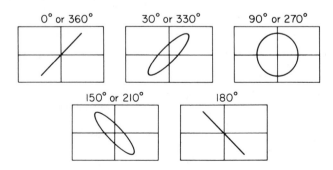

Fig. 5-37. Approximate phase of typical X-Y displays.

Once the oscilloscope's inherent phase shift has been established (Fig. 5-36a), and the op-amp phase shift measured (Fig. 5-36b), subtract the inherent phase difference from the phase angle to find the true phase difference.

For example, assume an inherent phase difference of 3°, and a display as shown in Fig. 5-36c, where A is 2 cm and B is 4 cm. The sine of the phase angle = A/B, or 2/4, or 0.5. From a table of sines, 0.5 = 30° phase angle. To adjust for the phase difference between X and Y channels in the oscilloscope, subtract the inherent phase factor (30° − 3°) for a true phase difference of 27°.

5-24. SLEW RATE

An easy way to observe and measure the slew rate of an op-amp is to measure the slope of the output waveform when a step input (squarewave or pulse) is applied. The input must have a rise time that exceeds the transient response of the op-amp. Thus, the output will not appear as a squarewave or pulse, but as an integrated wave. However, the input amplitude must be low enough to ensure that the op-amp will not be driven into saturation.

Slew rate can be measured either in the open-loop or closed-loop conditions. Generally, slew rate increases with gain. Thus, open-loop slew rate is typically higher than closed-loop slew rate.

It is important that the op-amp not be overdriven (in amplitude) when slew rate is measured. If the combination of input signal amplitude and gain (open or closed loop) is sufficient to drive the op-amp into saturation, and slew rate is calculated from the resultant squarewave at the output, the results will be uncertain. For example, assume that the op-amp has a slew rate of 30V/μS, but the op-amp saturates in 1 μS at 20V. It would appear that the slew rate is 20V/μS, since the output cannot rise to the full 30V.

The basic test connections and typical waveform for slew rate measurement are shown in Fig. 5-38. Note that the unity gain configuration is used. This provides closed-loop gain conditions, but prevents excessive output swing and saturation that might occur if the op-amp is set for high gain. If the input amplitude is kept within the normal differential limits of the op-amp, no saturation will occur. The input signal rise time should be in the order of 100 nS. This should be faster than any op-amp, and will result in an output wave with considerable slope, as shown in Fig. 5-38.

To find slew rate, adjust the input signal frequency until 1 μS portion of the output signal rise slope can be easily measured. If more convenient, use a one-half microsecond (500 nS) portion of the output waveform. Figure 5-38 shows slew rate measurement using a vertical scale factor of 10V/cm, and a horizontal factor of 500 nS/cm. Since the output rises about 20V from peak to peak in about 500 nS, the 1 μS rise must be about 40V. Thus, the slew rate is approximately 40V/μS. This may be listed simply as 40 on some datasheets.

5-25. OP-AMP TESTER

From a user's standpoint, there are four important op-amp tests: open-loop voltage gain, input offset voltage, output voltage limits

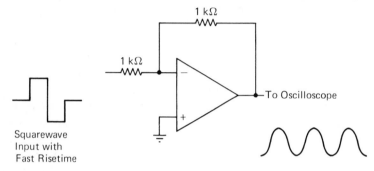

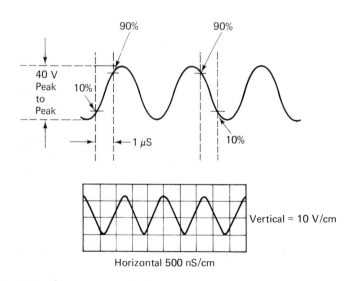

Fig. 5-38. Slew rate measurement.

and the transfer function (relation of output signal to input signal). Motorola has developed a basic circuit that will test these four primary functions. The Motorola tester (originally developed for IC op-amp tests) is simple and inexpensive, as is discussed in the following paragraphs. Also included in the following discussion is an elementary discussion of the parameters measured, and their relationship to closed-loop performance of the op-amp.

5-25.1 Test fixture

As with almost any tester, there must be a drive circuit, a power source, a device under test (DUT) and a display unit. These are shown in Fig. 5-39. The *X-Y* display function is provided by an oscilloscope, which is not part of the tester.

5-25.2 Power supply

The power supply shown in Fig. 5-40 is a straightforward shunt-Zener-regulated power supply. The pilot lamp is bridged across the filtering capacitors as an on-off indicator, and as a bleeder to remove the capacitor charge after power is turned off. The resistive divider in the transformer secondary serves as the input to the drive circuit.

5-25.3 Drive circuit

The drive circuit shown in Fig. 5-41 provides synchronized horizontal and vertical signals. In any test of this sort, it is essential that the vertical drive signal be synchronized with the horizontal sweep signal. The most straightforward way of accomplishing this is to drive the DUT input and the horizontal sweep with the same input.

Since a typical IC op-amp will have a possible gain of at least 1000, a 1000-to-1 precision resistive divider is placed at the input of the DUT. This brings into line the relative amplitudes of the *X* and *Y* inputs presented to the oscilloscope. For op-amps that show much greater gains, say greater than 100,000, another divider can be added. At this level, it is recommended that the divider be placed at the individual test socket to lessen the possibility of interference on the line.

It is desirable, although not necessary, that the retrace not appear on (or be blanked from) the visual presentation. This is accomplished by the drive circuit as a form of "intensity modulation." Note that this is not true intensity modulation, in which a signal is applied to the oscilloscope *Z*-axis, but produces the same effect. That is, the relative intensity

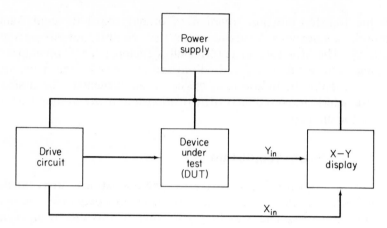

Fig. 5-39. Op-amp tester block diagram (Courtesy Motorola).

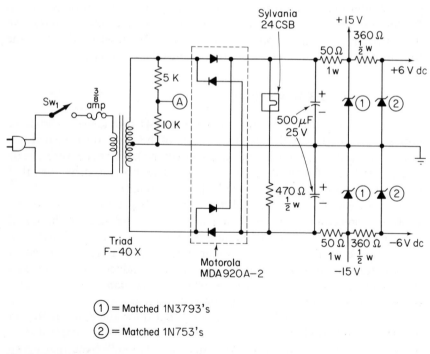

① = Matched 1N3793's

② = Matched 1N753's

Fig. 5-40. Power supply for op-amp tester (Courtesy Motorola).

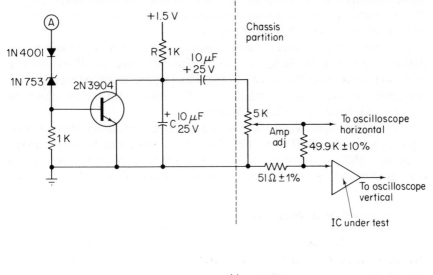

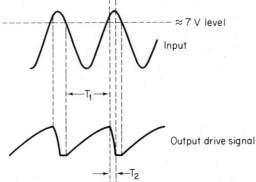

Fig. 5-41. Drive circuit for op-amp tester (Motorola).

of the oscilloscope trace is determined by the sweep rate, given a fixed set of other trace variables.

Operation of the drive circuit in Fig. 5-41 is simple. While the transistor is off, the RC combination charges toward the +15V supply. The time constant is chosen such that an ample amplitude results, while deleting the more exponential portion of the charge curve. The time constant must be sufficiently longer than the period of the reset rate (which in this circuit is 60 Hz).

A portion of the power transformer secondary voltage is used as the drive circuit input (Point A on Figs. 5-40 and 5-41). As the input sinewave exceeds a predetermined threshold voltage (approximately $2V_{BE} + V_Z$) the transistor is turned on, thus discharging the capacitor. This results in the waveform shown in Fig. 5-41.

The waveform is coupled to the op-amp and oscilloscope via the amplitude adjust potentiometer and the divider resistors. As shown by the Fig. 5-41 waveforms, the relative intensities of the trace and retrace are determined by the ratio of time T_1 to T_2. Thus, with the oscilloscope properly adjusted, the retrace "disappears."

The difference between the horizontal drive voltage and the op-amp input is a precise 1000-to-1. This permits the horizontal sweep voltage to be of sufficient amplitude to drive the oscilloscope, while retaining a low input to the op-amp to match realistic values of input offset voltage.

5-25.4 Tester wiring

The wiring of all tester sockets is shown in Fig. 5-42. Note that these socket pin arrangements are for Motorola IC op-amps. However, the socket connections can be rearranged to match other op-amps. In any event, the power supply terminals should be bypassed at each individual socket, and each socket should be provided with a proper frequency compensation network (capacitor, or capacitor and resistor, as necessary). Note that a single 2K resistor is used as a common load for all op-amps. The extact test load can be tailored individually, if desired. Also note that a switch is provided for three of the sockets. This is to test the hi-low gain option of certain Motorola IC op-amps.

Additional sockets for other op-amps can be incorporated as the demand warrants. Likewise, only one socket is needed if only one type of op-amp is to be tested. The maximum drive signal available is 8V peak-to-peak, open-circuit from the 5K amplitude adjust potentiometer. This should be sufficient to test most op-amps.

To add other sockets, it is necessary to install the power supply bypass and frequency compensation components, route the op-amp output to the oscilloscope vertical input, route the op-amp input to the drive circuit and adjust the drive voltage as needed. Keep in mind that the amplitude of the drive signal (to the op-amp) must be sufficient to overcome any input offset voltage, as well as to drive the op-amp into saturation.

5-25.5 Physical layout of tester

The exact physical layout of the tester is determined primarily by the number of sockets required to accommodate the various types of op-amps being tested. However, the following general notes should be observed for any tester layout.

Copper-clad laminate should be used for circuit wiring, if practical.
Ground loop pickup can be experienced if the power transformer is located near the low-level input leads. The power supply and drive circuitry should be located in a separate section of the tester, if practical. The dc leads and drive signal can then be routed to the sockets via feedthrough capacitors.

The 1000-to-1 divider should also be located near the sockets.

5-25.6 Interpreting the oscilloscope waveform

Figure 5-43 shows a typical oscilloscope waveform. Note that the op-amp under test is driven into saturation. The main features

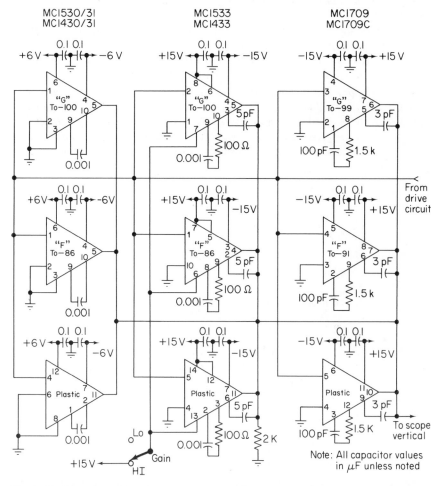

Fig. 5-42. IC socket wiring for op-amp tester (Courtesy Motorola).

of the transfer function are then used to find input offset voltage, open-loop gain and maximum output voltage. The following notes describe a typical operating procedure.

1. Ground the oscilloscope horizontal and vertical inputs temporarily.
2. Using the oscilloscope positioning controls, center the dot at the horizontal and vertical zero reference point (center of screen).
3. Insert the op-amp into the appropriate socket. Remove the ground from the oscilloscope inputs.
4. Increase the amplitude adjust potentiometer setting until the op-amp is in deep saturation (top and bottom of the output waveform flattened, as shown).
5. Read *maximum output voltage* (both plus and minus) directly from the transfer function trace. For example, assume that each vertical screen division equals one volt. The minus maximum voltage is then about 2.5V, whereas the plus maximum voltage is about 2.3V. Total output voltage swing is then about 4.8V peak-to-peak. That is, any output voltage greater than the 4.8V will be flattened as a result of saturation.
6. *Open-loop voltage* gain is equal to the calculated slope of the transfer function, multiplied by 1000. Any part of the transfer function can be used. However, the voltage required for deflection across *both horizontal and vertical* screen divisions must be known. For example, assume that both horizontal and vertical screen divisions equal one volt. Under these

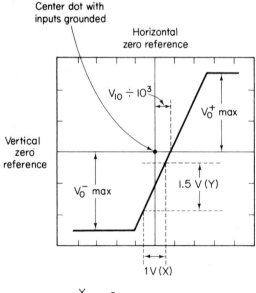

$$A_{vol} = \frac{Y}{X} \times 10^3$$

Fig. 5-43. Interpreting oscilloscope waveforms of op-amp tester (Courtesy Motorola).

conditions, the X dimension of the transfer function sampled in Fig. 5-43 is equal to 1V, whereas the Y dimension is equal to approximately 1.5V. Thus, $X/Y = 1.5$, and the open-loop gain $= 1500$ (1.5×1000). If the oscilloscope horizontal and vertical screen division dimensions are identical (say both divisions equal 1 cm), the exact voltage need not be considered. Instead, only the ratio of Y/X for any given portion of the transfer function is used.

7. *Input offset voltage* is equal to the horizontal displacement from the horizontal zero reference, to the point where the function crosses the vertical zero reference, divided by 1000. For example, assume that each screen division equals one volt. Under these conditions, the function crosses the vertical zero reference about 0.7V from the horizontal zero reference in Fig. 5-43. Thus, input offset voltage $= 0.7$ mV ($0.7/1000$).

8. A measure of op-amp linearity can also be obtained by noting the linearity of the transfer function between saturation points. For example, if the line is perfectly straight, the op-amp is perfectly linear.

9. In reading open-loop gain and input offset voltage from the oscilloscope presentation, it is advisable to increase the horizontal sensitivity so that resolution is increased for better accuracy. Keep in mind that if the oscilloscope voltage ranges are changed, it may be necessary to check the zero references (both horizontal and vertical). However, it is usually not necessary to check the zero reference when testing one op-amp after another.

5-25.7 Analyzing measured parameters

Open-loop voltage gain can be defined as the ratio of a change in output voltage to a change in input voltage. The ideal op-amp has an infinitely high open-loop gain, since the primary function is to amplify. In general, the higher the gain the better the accuracy. The significance of open-loop gain is many times misapplied. From a user's standpoint, open-loop gain determines closed-loop *accuracy limits,* rather than the ultimate accuracy.

Referring to Fig. 5-44 (an ideal op-amp), the closed-loop gain of the circuit is:

$$\frac{e_o}{e_{in}} = -\frac{A_{VOL}\left(\dfrac{R_2}{R_1 + R_2}\right)}{1 + A_{VOL}\left(\dfrac{R_1}{R_1 + R_2}\right)} \tag{1}$$

If the op-amp in Fig. 5-44 shows infinite open-loop gain, equation 1 reduces to the ratio of the two passive feedback elements, which is the ideal closed-loop gain of an op-amp connected as shown in Fig. 5-44:

$$\frac{e_o}{e_{in}} = -\frac{R_2}{R_1} \qquad (2)$$

The error in the closed-loop gain ($error_{CL}$) of an op-amp may be represented as:

$$\%(error_{CL}) = \frac{\left(\dfrac{e_o}{e_{in}\ IDEAL}\right) - \left(\dfrac{e_o}{e_{in}\ ACTUAL}\right)}{\dfrac{e_o}{e_{in}\ IDEAL}} \times 100 \qquad (3)$$

which, after insertion of equations 1 and 2, reduces to:

$$\%(error_{CL}) = \frac{100}{1 + A_{VOL}\left(\dfrac{R_1}{R_1 + R_2}\right)} \qquad (4)$$

The closed-loop gain error is a direct function of the loop gain

$$A_{VOL}\,\frac{R_1}{R_1 + R_2}$$

rather than solely open-loop gain. Open-loop gain is the limiting factor in closed-loop gain error; loop gain establishes the accuracy.

Output voltage swing (V_o max) can be defined as the peak output voltage swing, referred to zero, that can be obtained without clipping (due to saturation). A symmetrical voltage swing is generally implied. However, if V_o^+ max and V_o^- max differ, the maximum symmetrical voltage swing is limited by the *lesser absolute value*. Output impedance, load current and frequency directly affect V_o max.

In addition to the limiting factor of V_O max on the output swing, the transfer linearity affects the maximum output swing, within distortion limits. The deviation of the transfer function from a perfect straight line, within the saturation limits, indicates the degree of distortion that can

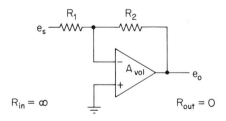

Fig. 5-44. Idealized operational amplifier (Courtesy Motorola).

be expected in the output signal, as well as the peak voltage at which this distortion occurs.

Input offset voltage (V_{io}) can be defined as that voltage that must be applied at the input terminals to obtain zero output voltage. V_{io} also indicates matching tolerances in the differential amplifier stages of the op-amp. V_{io} is primarily determined by the base-emitter voltage (or gate-source voltage) of the op-amp input stage, and unbalance in the second stage, attenuated by the gain of the first stage.

In general, V_{io} will be the major source of offset voltage error in low source impedance circuits. Op-amps with minimum V_{io} are better matched, and will generally track well with temperature variations. Thus, such op-amps will show minimum output drift with temperature variations.

A factor not always considered when determining the contribution of V_{io} in closed-loop operation is that the error is not simply increased by the ratio of feedback resistance to input summing resistance but by unity (or 1) plus this ratio. At high closed-loop gain levels, the difference is of little concern. However, at unity gain operation, the difference is a factor of 2.

INDEX